Space and time in the modern universe

Space and time in the modern universe

P. C. W. DAVIES

Lecturer in Applied Mathematics, King's College, London

CAMBRIDGE UNIVERSITY PRESS

CAMBRIDGE

LONDON · NEW YORK · MELBOURNE

Published by the Syndics of the Cambridge University Press
The Pitt Building, Trumpington Street, Cambridge CB2 1RP
Bentley House, 200 Euston Road, London NW1 2DB
32 East 57th Street, New York, NY 10022, USA
296 Beaconsfield Parade, Middle Park, Melbourne 3206, Australia

First published 1977
Reprinted 1977,1978

Filmset by Cox & Wyman Ltd, London, Fakenham and Reading
Reprinted in the United States of America
by R.R. Donnelley and Sons Co. Crawfordsville, Ind.

Library of Congress Cataloguing in Publication Data

Davies, P. C. W.
Space and time in the modern universe.

1. Space and time. I. Title.
QC173.59.S65D39 530.1'1 76-27902
ISBN 0 521 21445 9
ISBN 0 521 29151 8 pbk.

CONTENTS

PREFACE

The structure of space and time lies at the very foundation of both physical science and our perceptual experience of the world. They are concepts so fundamental that in everyday life we do not question their properties. Yet modern science has discovered situations in which space and time can change their character so drastically that remarkable and unexpected phenomena occur. Many of these situations owe their appearance to recent developments in astronomy. The possibilities of the existence of black holes or a big bang origin of the universe have stimulated detailed investigations into the behaviour of space, time and matter, when gravity becomes overwhelmingly strong. The results indicate that space–time itself may collapse out of existence under some circumstances. The consequences of these recent developments for the nature and evolution of the universe are far-reaching and profound.

This book sets out to examine the exciting and sometimes enigmatic discoveries about space, time and the universe made by scientists in recent years. Any proper understanding of these modern advances depends on the reader having a grasp of the physicists' new model of space–time structure.

The changing nature of this model over the last 150 years, and our evolving perspective of the universe that accompanies it, are carefully described in the forthcoming chapters. The reader will discover how the space–time picture of Newton, so close to that of our familiar experience, has been repeatedly remodelled and modified to accommodate new physical theories. The laws of electromagnetism, with the abortive invention of an ether, the strange four-dimensional space–time of the theory of special relativity, the intriguing notion of curved space in the general theory of relativity, not to mention the sheer outlandishness of the quantum theory, which predicts that black holes may evaporate away to nothing and space–time may break apart below microscopic dimensions – all have demanded a fresh conceptual structure for these most primitive of all physical entities.

Woven into this changing picture of space–time structure is a parallel story of confusion and paradox concerning the asymmetry of the world between past and future. The nature and origin of this time asymmetry reside in the basic laws of thermodynamics and the conditions placed on the universe at its very beginning. A study of time asymmetry allows us to choose between two distinct models of the universe: one which slowly runs down to cold, featureless sterility, or one which undergoes rejuvenation, time-reversal, or simply crushes itself out of existence.

So much of the subject matter discussed overlaps with domains of human intellectual activity normally associated more with religion and metaphysics than hard science. Yet today, science is on the brink of suggesting answers to many of the questions so long puzzling to theologians and philosophers alike. No account of these developments is complete therefore, without an examination of the place of mankind and human society in the new universe. The impact on society of changing ideas about space, time and the nature of the cosmos has always been profound. The revolution now in progress could alter forever mankind's perspective not only of the universe, but also of our own place in that universe.

I have intended this book to be both educative and entertaining, for that is the experience of the researchers and teachers of the subjects expounded. Unravelling the scientific mysteries of our world is one of the most satisfying of human endeavours, and I have attempted to communicate the sense of excitement and awe felt by the practitioners of modern science in these challenging times.

I have not assumed a high level of expertise from the reader. Scientifically inclined laymen with no mathematics and a sprinkling of physics should have little difficulty in following much of the discussion. However, because a lot of the material is very advanced, students of physics, astronomy, applied mathematics and philosophy will find a wealth of information relevant to their lecture courses or research.

Finally, I should like to thank my colleagues at King's College, London, whose opinions and observations have helped shape my own view of space–time physics and cosmology.

Note on nomenclature. Physicists and astronomers often have to deal with very large or very small numbers, and it is usually inconvenient to write them out in full. For that reason the shorthand 'powers of ten' notation will frequently be used in this book. In this system, the number one followed by *n* noughts is simply written 10^n. Thus one thousand is 10^3, one million is 10^6 and one billion, here taken as the USA billion (one thousand million) is 10^9. Small numbers are similarly represented using negative powers (10^{-n}). For example, one-thousandth is 10^{-3}, one-millionth is 10^{-6} and one-billionth is 10^{-9}.

King's College, London P. C. W. Davies

1 The many faces of space and time

1.1 General concepts

Space and *time* are two very overworked and ambiguous words in the English language. In common parlance space is identified with emptiness, extension, volume – room to put things in. Modern jargon often refers to 'outer space'. This is the region beyond the Earth which is imagined as a total void. Strictly speaking, outer space is not a perfect vacuum. The vast gaps between stars and planets always contain at least a minute amount of matter, and a considerable amount of radiation of one sort or another. Nevertheless, the word space conjures up a picture of *emptiness* – what is left when all tangible things have been removed. Consequently, most people think of space as a container or arena, with the universe – all the galaxies, stars, planets – contained in it. Space does not go away when matter is present, it gets 'filled up'.

This view of space as the *absence* of things makes it hard for many people to understand why scientists should want to make theories about it. After all, if space is nothing, then there is nothing to say about it!

The scientists' view of space is quite different. First, to dispel a possible misconception, scientific theories of space are not theories about outer space. The properties of space beyond the Earth are almost everywhere closely similar to the properties of space at the Earth's surface. When Newton and Leibniz conjectured about the nature of space, they knew little of modern astronomy.

The modern scientist regards space as possessing many levels of structure. Indeed, certain branches of modern physics suggest that material objects are really just a minor disturbance on this underlying structure. Rather than picturing the universe as contained *in* space, modern cosmology regards both material objects *and* space as together constituting the universe. The universe is space and matter.

Space therefore stands alongside matter in possessing physical status, properties and structure. Much of this structure was familiar to the early Greeks, who systematically recorded it in their axioms and theorems of geometry. Much later, Isaac Newton (English, 1642–1727) discovered further properties through his study of *motion* – the dynamics of bodies moving through space. Newton regarded space as a substance which could *act* on material bodies dynamically.

In contrast to the picture of space as a physical entity, which can exist in its own right independently of matter, there is a long tradition among certain scientists and philosophers of attempting to reduce all the properties of space to relations between material bodies. The rationale behind this relationist school of thought is based on the fact that the acquisition of information about space depends upon measurements and observations carried out using material objects, light signals and so on. The relationist views space simply as a linguistic convenience, a means of expressing relations. The spatial relationships between material bodies are regarded as no more requiring the existence of a special physical substance called 'space' than the relationship between Englishmen requires a physical substance called 'citizenship'. In later sections it will be discussed how the relationist school has fared with the development of physics through the last three centuries.

Many of the properties which are attributed to space (or to the relationship between objects) are well known to most people and usually taken for granted. Other properties are far more subtle and are only known to physicists or mathematicians. The complexity and richness of this structure is soon revealed when the properties of real physical space are compared with mathematical models of spaces in which some of the structure is absent. The next section describes the modern mathematician's description of physical space, and a measure of the complex nature of real space is the amount of mathematical concepts that have to be used in order to describe it adequately. Before these mathematical models are discussed, some consideration must be given to the use of the word *time*.

The human experience of time is fundamentally different from that of space. In a sense time is the most elementary aspect of all experience. It enters into our consciousness directly and frames our perceptions, attitudes and language. Unlike space, the structure of which is only appreciated by observation, and abstraction away from the familiar, time is perceived to possess structure at a most fundamental level. The acquisition of information about space is obtained through the laboratory and the external senses. The acquisition of information about time has an additional 'back door' into our minds. The structure which is perceived through this back door may be described as a flow or flux, a passage from past to future sweeping our conscious experience along from one present moment to the next. In the popular mind space is empty, but time is full of activity.

Once again the scientific picture of time differs radically. It may not seem very obvious that time and space should be coupled together in any fundamental way, because they are such totally different human experiences. But the mathematician's description of time turns out to be very similar to that of space. In addition, space and time are linked together by motion, and it arises from the study of the motion of material bodies and light signals that space and time are actually two aspects of a single unified structure called space–time.

It is one of the most perplexing puzzles in physics that the elementary conscious experience of time – the flow or motion of the present moment – is absent from the physicist's description of the objective world. Whether this is due to a deficiency in the framework of physics, which pays scant attention to the role of the conscious mind in the universe, or whether it is because the passage of time is an illusion, is by no means clear. Nevertheless, because of this profound awareness of time by human beings, the acts of violence performed on our intuitive picture of time by modern theories such as relativity are often far more disturbing than the similar savagery carried out on space. Deep philosophical issues and paradoxes invade temporal speculation, cutting across issues such as freewill and death. The interaction of the physical world of the scientist and the metaphysical world of our minds leads to some strange and uneasy conflicts.

1.2 Mathematical models of space

A theory of space, like all theories in science, requires a *model.* In common with most good physical models, it should have a mathematical description. In order to construct a model which bears a good resemblance to space in the real world, it is necessary to incorporate quite a large number of mathematical concepts. The reader should not be perturbed by this, because only an elementary descriptive account of these concepts is needed to understand the rest of this book. An appreciation of the concepts reveals just how specific the nature of our real space is.

Mathematicians use the word space to mean any collection of points. A point is the primitive object in a description of mathematical model spaces and can be pictured as the limit of a small circle as the radius of the circle is shrunk to zero. Points therefore have no size, extension or interior. Any structure in the space is imposed on the collection of points, not on the individual points.

A mathematical model space, it must be emphasised, can have a variety of purposes. It may be used in the description or solution of various types of problem from other branches of mathematics, or simply for its own intrinsic interest. Many abstract mathematical spaces are used in everyday life, such as plotting a graph. The sheet of paper on which the graph is drawn is a set of points, and the graph itself is a subset depicting some sort of relationship, like the variation in the national balance of payments with time. A mathematical space may also be used as a model for real, physical space. Clearly there is more to real space than just a collection of points. Several levels of progressively more complex descriptive structure need to be imposed on this collection before the familiar properties of real space emerge, and perhaps still further structure to produce an adequate description of some of the strange properties which modern physics has uncovered.

In this section the various levels of descriptive sophistication which must be imposed on the collection of points in order to arrive at a plausible model of real space will be itemised and briefly discussed. Just which features of real space should be mimicked in the model obviously depends on which theory of space is being discussed. However, there are certain basic features which most theories have in common, and these will now be described.

(*a*) *Continuity.* It is generally supposed that any interval of space may be subdivided again and again without limit. This is purely an act of faith, because it has not yet proved possible to find out what is happening inside distances much less than 10^{-13} cm. Nevertheless, the assumption of infinite divisibility is nearly always made, so that we must envisage space as an infinite collection of points packed so closely together that they are *continuous.* This description is necessarily heuristic only, because continuity is a fairly sophisticated concept which has been well understood by mathematicians only in the last century or so. Nevertheless, it seems natural that a continuous line has in some sense *more* points than an unending row of disconnected points (like that shown in fig. 1.1) even though in both cases the number is infinite. This difference is often characterised by pointing out that the lattice shown in fig. 1.1 may be labelled with the integers 1, 2, 3,... but the points on the continuous line cannot; they require as well all the numbers in between the integers (e.g. 1.5321648...) in order to make the labelling complete. From time to time a lattice space such as that shown in fig. 1.1 is suggested as a model for real space. We shall not dwell on the properties of such unconventional models.

Fig. 1.1. The notion of a continuum. Set *A*, when extended for ever to the right (or left) contains an infinity of points which may be labelled with the infinity of whole numbers 1, 2, 3...Set *B*, the line segment, also contains an infinity of points, even when the line is finite in length, but they are so closely packed that there are no 'gaps' between them; the line is *continuous.* The difference is not merely one of *scale.* There are not enough whole numbers to label the points on the line. *B* actually contains *more* points than *A*.

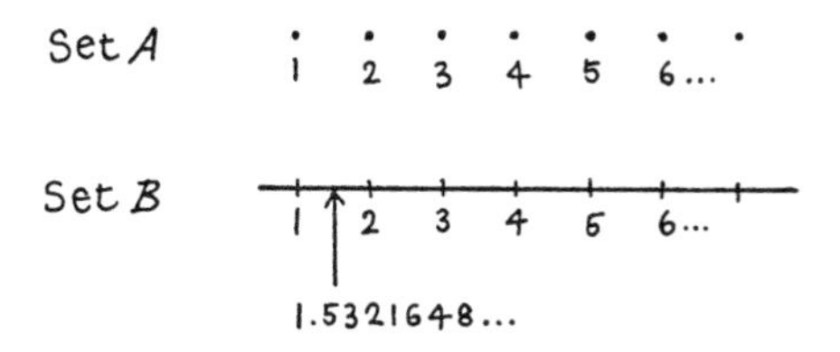

In a continuous space, or *continuum*, every point has a neighbourhood which, however small it is shrunk, still contains an infinite number of other points. Furthermore, we must require that around any two distinct points, disjoint (i.e. non-overlapping) neighbourhoods may be constructed.

(*b*) *Dimensionality*. A well-known specific feature of real space to be imposed on this continuum is the frequently stated fact that it is *three-dimensional*. The easiest way to understand the meaning of this is to start with a point, which is defined to have dimension zero. Points can then be used to form the *boundary* of a space of dimension one. Consider for example a finite straight line. The boundary of the line is the two end points. The line

Fig. 1.2. Boundaries and dimensionality. The straight line is a one-dimensional space with two boundary points *A* and *B*. Any neighbourhood of these points (marked by ( )), however small, always contains an infinity of points both from the space and not from the space. In contrast, the circle is an unbounded one-dimensional space; none of its points possess the above property.
Just as the zero-dimensional points *A* and *B* form a boundary for the straight line, so the one-dimensional circle is a boundary for the two-dimensional disk. (A typical boundary point *C* and its neighbourhood are shown.) In contrast the surface of the sphere has no boundary points; it is an unbounded two-dimensional space.
These considerations may be extended to any number of dimensions.

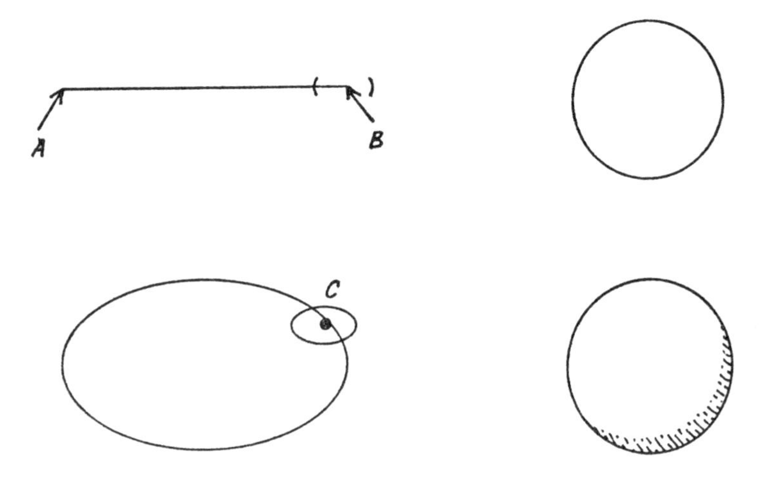

(of dimension one) in turn may be used to form the boundary of a space of dimension two, e.g. the one-dimensional circle bounds the two-dimensional disk. Then a two-dimensional surface may be used as a boundary to a three-dimensional volume, and so on. The statement that space is three-dimensional therefore refers to the order of progression along this series. There is no limit *mathematically* to the number of dimensions a space may possess. Indeed, an important branch of mathematics with applications in physics deals with infinite dimensional spaces! It is not known why real space is three-dimensional. It is interesting to consider

Fig. 1.3. Disconnected and connected two-dimensional spaces. Points such as *P* and *Q* in the space drawn at the top cannot be connected by a continuous line which remains totally within the space. This is a disconnected space. In contrast, all parts of the torus and the sphere are connected, but in different ways. On the surface of the sphere, circles such as *C* may always be shrunk to a point. In the case of the torus this is true for *C* but not *D*. The sphere is called simply connected, the torus multiply connected.

A flat inhabitant of these connected spaces could easily deduce their differences solely from observations within the two-dimensional surfaces. It would not be necessary for the flatlander to leave the surface and see in three dimensions (as we can) how the torus or sphere are embedded in three-dimensional space, in order to deduce whether or not they were simply or multiply connected, bounded or unbounded. Similar remarks apply to our own three-dimensional universe.

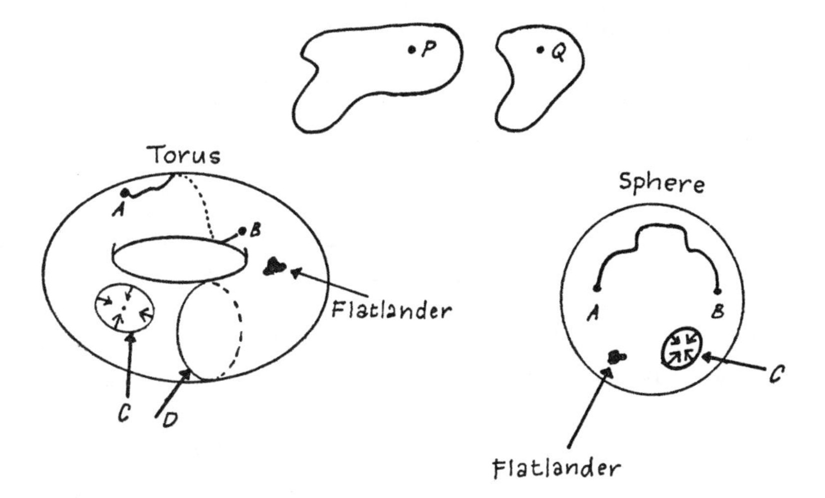

the properties of worlds in which space is say two-, or six-dimensional. Sometimes things such as wave propagation or electric phenomena are very different in these imaginary worlds.

(*c*) *Connectivity*. There is no reason why real space should not consist of a number of disconnected pieces. However, we could not know of a region of space disconnected from ours so we shall not pursue this. Nevertheless, even a single space may be connected together in many different ways. For example both the surface of the torus (a doughnut shape) and the surface of the sphere (see fig. 1.3) are connected spaces in the sense that every point may be joined to every other by a continuous curve through the space; however, they are clearly connected differently. One way of expressing this is to note that a simple closed curve (such as a circle) may always be shrunk continuously to a point on the surface of a sphere, but not necessarily on the torus. We do not know if our universe is like the surface of the sphere, or the torus, or some more complicated system. However, in the region of the universe that we actually observe, it is simply connected, like the surface of the sphere.

At this stage the reader may well be suffering from the usual mystification of how we can meaningfully discuss real space being like a torus, or consisting of disconnected pieces. If this were so, what would be 'outside' space, filling the hole in the middle of the torus, etc? It is one thing to discuss mathematical spaces in which two-dimensional surfaces can be wrapped around into a torus, but surely this can only be done by embedding (i.e. enveloping) the torus in real three-dimensional space? What 'superspace' is real space to be enveloped in? Problems of this sort always cause a certain amount of intellectual difficulty to the non-mathematician.

A space is defined by its properties. Many of these properties may be made quite precise without any reference to an embedding space surrounding the space of interest. For example, a two-dimensional 'flatlander' could deduce that he lived on the surface of a torus solely by observation from *within that surface* (say by testing whether all circles could be shrunk to a point). Mathematically, there is no difficulty in extending the discussion of a two-dimensional toroidal surface to a three-dimensional toroidal volume without introducing an embedding 'superspace'. Nevertheless, it is helpful sometimes to imagine a higher dimensional embedding space just for intuitional convenience, but one should not expect any discussion of the nature of this embedding space, for it is only an artifice.

(*d*) *Orientability.* With this caution in mind, we shall frequently describe the properties of real three-dimensional space by analogous two-dimensional models embedded in three-dimensional space, for clarity of exposition. For example, such an analogy is helpful in the discussion of another important property which is usually assumed to hold for real space; that of *orientability.* It is a familiar experience that a left-handed glove cannot be turned into a right-handed glove no matter how it is twisted and turned (except inside out). In addition, it would generally be assumed that even if one glove was taken to a distant region of the universe and returned it would not change its handedness, i.e. a left-hand glove would not come back as a right-hand glove. But mathematicians frequently encounter spaces in which this change of orientation does take place. One such example is the Möbius strip (after August F. Möbius, German astronomer and mathematician, 1790–1868), which is a two-dimensional space conveniently described by drawing it embedded in three-dimensional space as shown in fig. 1.4. It consists of a strip with a single twist in it, and a moment's thought will show that a left-

Fig. 1.4. Non-orientable space. The Möbius strip has the strange property that a right-handed glove changes into a left-handed glove when transported once around the strip. (No distinction is to be made between the front and reverse surfaces of the strip.)

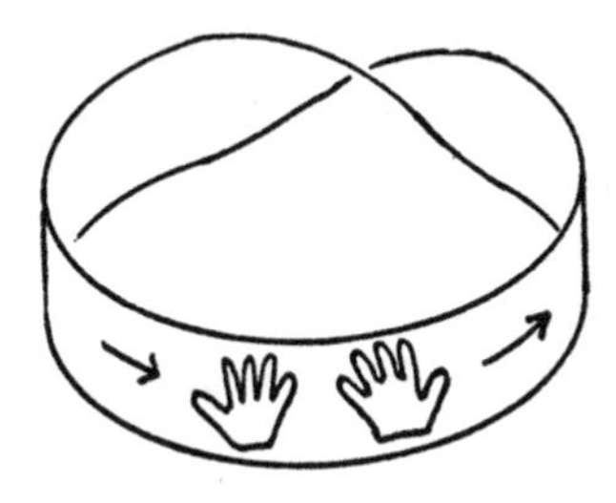

handed glove turns into a right-handed glove when transported in a closed curve round the strip. (These are two-dimensional gloves of course, so we do not distinguish a 'front' and 'back'.) Analogous three-dimensional 'Möbius' spaces can be described mathematically in which this restriction is removed. The Möbius strip is a non-orientable space. There is no evidence that our universe is non-orientable.

The features (*a*)–(*d*) which have been discussed above are known as *topological* features; their study is the subject of *topology*. They depend only on the *continuity* of the space and not on properties such as size or precise shape. Already real space is seen to possess a large amount of structure beyond being just a 'set of points'. There is continuity, dimensionality, connectedness, orientability and a number of other mathematical properties beyond the scope of this book.

But even with these restrictions, we may construct mathematical spaces with wildly different properties from real space. Further important restrictions must be imposed before we have useful models of the real universe. One of the most obvious practical properties of space is the way in which points can be located by continuous labels or coordinates. To take a familiar example, a town may be located by specifying its latitude and longitude, two numbers which label points continuously on the two-dimensional surface of the Earth. We could construct a system of three numbers to locate objects anywhere in space, e.g. latitude, longitude, altitude. Naturally the *value* given to these numbers depends on the type of coordinate system used. For example, shifting the line of zero longitude from Greenwich to Paris changes one of the two numbers locating a town. Or suppose that we choose to locate all points on Earth by specifying their direction and distance from Mecca, instead of using latitude and longitude? It may also be necessary to use more than one different coordinate system to cover the whole space properly. For example, latitude and longitude break down near the poles (at the north pole all directions face south). We must then require that there is a sensible relation between one coordinate system and the other in the regions where the two systems overlap. A space which carries consistent continuous coordinates is called a *manifold.*

In addition to being a manifold, real space has *geometrical* structure. For example, there exists a shortest path between any two points. Furthermore, *distances* and *angles* may be defined. Spaces with these features are called *metric spaces*. There are many different kinds of metric space. Before 1915 it was assumed (except by a few mathematicians) that the real universe was a metric space which was restricted to obey the rules of *Euclidean geometry*, named after the Greek geometer Euclid. In this system the angles of a triangle always add to 180° and parallel lines can always be drawn. This is just the familiar geometry of our school-days so the reader need not worry yet about this. As we shall see, modern theories of space allow the remarkable property that the metric structure may vary from place to place and time to time, and the rules of Euclid no longer apply.

Before the discussion of the mathematical properties of real space is ended, a word must be said about time and space–time. Clearly time may be assumed to share many of the properties of space. For instance, topological properties such as continuity, connectedness and orientability are probably the same, though time is only one-dimensional rather than three. It also has a metric structure because we can define the distance between two points in time as the interval between two events (e.g. one o'clock to two o'clock). For these reasons time can be regarded as a one-dimensional mathematical metric *space*, which should not be allowed to confuse the reader into thinking that time is really space in disguise or something. In addition, it has proved more accurate to regard the three dimensions of space and one dimension of time as a unified four-dimensional space–time, which is also a metric space. Consequently the word 'space' will often be used in this mathematical context to cover aspects of both physical space and time, or space–time.

1.3 Newtonian space and time

The metric properties of space near the surface of the Earth were explored in great detail by the early Greek geometers, and these *static* features of the world became formalised in the axioms and theorems of Euclidean geometry. But the *dynamical* properties of the world were not incorporated into a systematic mathematical theory until the work of Isaac Newton in the seventeenth century. Newton presented a *theory of motion* of material bodies. Because the path of a moving body is *through* space *in* time, this theory necessarily connects space and time together in a set of laws. Newton discovered simple mathematical relationships governing the motion of idealised rigid bodies. The impact of his monumental work shaped the structure of physical science for centuries after.

The model of space which Newton proposed was that of a *substance*, with independent existence, through which material bodies and radiation move, rather like fish swimming through water. Thus every object has a unique position and orientation *in* space, and the distance between two events is well defined, even if the events occur at different times.

Newton's concept of time relied heavily on the notion of *simultaneity*. Time in this model is universal and absolute. A universal time attributes meaning to the notion of events being simultaneous (i.e. occurring at the *same* time) – when these events are at separate points in space. Thus twelve o'clock in London is twelve o'clock in the whole universe (though it may be called seven o'clock in New York, but that is mere convention, it being the *same moment* according to Newton's theory). Newtonian space and time are supposed to remain absolute – a fixed arena or framework, unchanged by the behaviour of the contents (material bodies). For although, as we shall see, space is taken by Newton to act on matter under some circumstances, matter does not react back on space.

By treating space as an independent substance, Newton's theory comes into conflict with the relationist school, who believe that spatial and temporal discussion is merely a linguistic convenience for describing the relationships between material bodies. According to this point of view, to treat space as a physical entity is as absurd as treating the 'bad atmosphere' after a quarrel as an actual physical substance. For to say that someone is embarrassed by a bad atmosphere after a quarrel is merely a linguistic convenience of saying that they are embarrassed by the mood of the combatants after the quarrel. No one suggests that this mysterious 'atmosphere' exists independently of the combatants, to be detected with an instrument perhaps! Space as a physical object only makes sense if it can be detected, or if it can exert physical influences. How, for example, are we to determine the *position* of an object in space? Space is, by definition, without any features or landmarks. Naturally we may determine the position of an object *relative* to some other set of objects; for example, a specification of latitude and longitude fixes the distance of a place from the equator and the Greenwich meridian respectively. Moreover, all the geometrical properties of space are only inferred from observations with material objects and light signals; for example, it is easily verified that to a high degree of approximation the angles of a triangle add up to 180° if we equip ourselves with theodolites and poles. How could this property be deduced in empty space? However, one part of the ocean looks much the same as the other, but its existence as a separate material entity is not in doubt because we may travel *through* the ocean and feel its resistance. Does the motion of an object through space lead to detectable effects? Can space act on a moving body in the way that the sea acts on a moving fish?

In Newton's model of space and time, it is meaningful to discuss the *velocity* of an object through space. The question 'how fast are you moving?' is a common enough phrase for which one usually expects a sensible reply. A man sitting in his living-room would normally regard himself at rest. Yet a moment's reflection recalls that he is really travelling with the Earth around the sun. So how fast is the Earth moving? This question cannot be answered without knowing how fast the sun is moving. In fact, the sun is in rotation around the galaxy. Nor can we stop there, for all known galaxies are receding from one another in a general pattern of expansion. The universe is full of motion. These considerations cast doubt upon any mechanical means for deciding whether *anything* is at rest in the universe. How is such a state of rest to be determined?

For many centuries it was believed that the Earth was at rest in the universe. The sun, moon and stars revolved around the Earth with clockwork precision, but the Earth itself was fixed.

Nicholas Copernicus (Polish, 1473–1543) destroyed this cosy image of mankind at the centre of the universe (and of God's attention) by demonstrating that the sun is at the centre of the solar system and the Earth revolves around it. Mankind has never quite recovered from the intellectual shock of losing this privileged status of the Earth.

Small children find it difficult to accept the fact that the Earth moves, because it does not *feel* as though it is moving. An important insight can be gained into the nature of mechanics by analysing just what type of motion *is* felt. An aeroplane passenger, when wishing to determine whether or not he is flying through the air or is stationary on the ground has merely to look out of the window for an answer. If the aeroplane had no windows he would be less sure of this, but occasional bumps from turbulence, or a banking or climbing manoeuvre by the pilot would be sufficient to convince him that he was airborne and moving. However, if the flight were very smooth, even these sensations of movement would be lost. Indeed, it is very easy to be fooled psychologically about one's state of motion. The reader will no doubt have had the common experience of sitting in a railway carriage which appears to be pulling out from the station, only to discover a few moments later that the carriage has not started off at all, but another train on a parallel track has departed in the opposite direction.

A generalisation from experiences such as these reveals that the sensation of motion only occurs when the motion is *not uniform*. For example, disturbing bumps in an aircraft, or a change of speed, height or direction can all be felt by a blind passenger riding inside the aircraft. Likewise, the tell-tale lurch of a train which brakes suddenly, or pulls away sharply is sufficient to convince a passenger whether it is his own, or his neighbour's train which is leaving the station. Stated more precisely, motion with a uniform velocity (no change in speed or direction) cannot be felt, but *accelerated* motion, where the velocity *changes* either in magnitude or direction (this includes, of course, deceleration) is readily felt.

Newton placed these rather anthropomorphic considerations on to a secure scientific basis by formulating his laws of motion in such a way as to be quite independent of the velocity of a physical system, and dependent instead only upon its acceleration. He was thereby proposing that if two systems are both in uniform motion with different velocities, it cannot be established by any experiment whatever whether one of them is really at rest and the other moving (or vice versa) or that they are both moving. All that can meaningfully be said about the two systems is that they are in uniform motion *relative* to each other. It is usual to refer to a state of motion as a *reference frame*, and we may think of a hypothetical observer attached to each reference frame. Newton's laws then deny the existence of a privileged class of reference frames which may be described as 'at rest'. In Newtonian mechanics *all uniform motion is relative.* If we say that a car is moving at 50 kilometres per hour we simply mean by this '50 kilometres per hour relative to the pavement'. Newton's laws imply that that is *all* such a statement can mean.

In sharp contrast to the relativity of uniform motion, accelerated motion is *absolute* in Newton's theory. That is, we may devise experiments to answer quite unambiguously the question: 'is this reference frame accelerating (through absolute space)?', and these experiments may be carried out entirely from within the accelerated system without consulting the outside world. To return to one of the above-mentioned examples, an egg placed

Fig. 1.5. Uniform velocity is relative. Two space capsules approach each other at 10 000 kilometres per hour in remote space. Each occupant feels no sensation of motion and regards his capsule as at rest while the other is in motion. Who is right? The question cannot be answered. No mechanical instrument whatever carried by the capsule could detect its uniform velocity. Only the relative velocity between the two is observable.

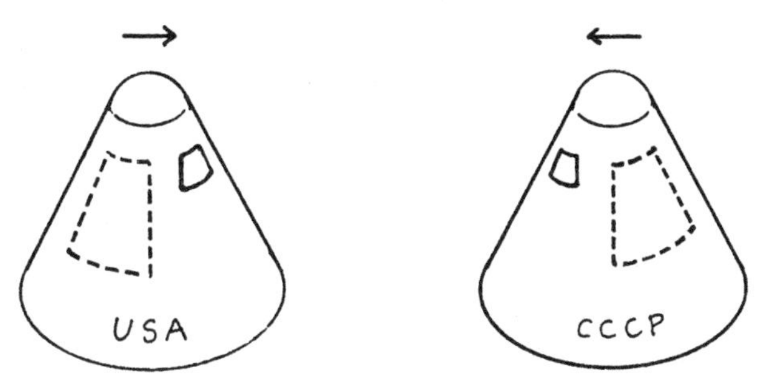

on a flat table in an aeroplane flying steadily will do nothing extraordinary to betray the uniform motion of the plane. However, if the plane brakes suddenly or banks, the egg will roll off the table and break. Accelerations break eggs, uniform motions do not.

From Newton's laws of mechanics, uniform motion emerges as a rather unremarkable condition. Consequently, Newton made no attempt to *explain* uniform motion, which may be regarded as a *natural* state, but he instead proposed that accelerated motion always requires a cause and he called these causes *forces*. For example, a stone falls to the ground because the force of gravity accelerates it downwards. A similar consideration applies to the motion of the Earth around the sun. Newton's laws assert that the *velocity* of our planet in space requires no explanation, and this is fortunate for when we look along the direction of the Earth's velocity, not only is there nothing remark-

Fig. 1.6. Circular motion is acceleration. The Earth moves around the sun in a nearly circular orbit at constant speed. However, the *direction* of the Earth's velocity changes continually. When at *A* the Earth is moving towards a point *B* in space, but because of the curvature of the orbit it actually swings inwards to the point *C*. The *change* in the diıection of motion, along *BC*, is towards the sun. This change in the velocity – i.e. the acceleration – is caused by the gravitational force of the sun pulling the Earth along the direction *BC* (double arrow). There is no force acting along the direction of the orbital velocity (single arrows).

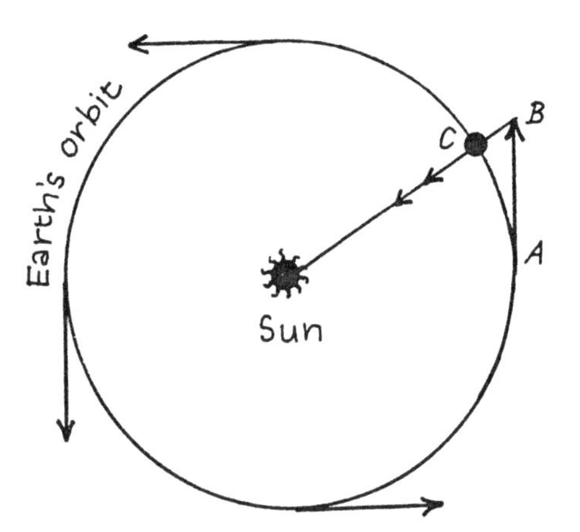

able in sight which might account for this velocity, but also the direction of the motion changes continuously as the Earth swings in a curved path around the solar system. However, it is just this curvature of the Earth's orbit which does require explanation in the theory, for a body which moves in a curved path is always accelerating in a direction askew to the path. For instance a planet which moves in a circle is always having the direction of its speed changed towards the centre of the circle (see fig. 1.6). If then the Earth is always accelerating towards the centre of its (nearly circular) orbit, it is there that we must look for an explanation of its behaviour, rather than along the direction of the motion. And when we do look to the centre of the Earth's orbit we indeed see something remarkable – the sun. It is the gravitation of the sun which forces the moving Earth to curve around it forever. If the sun were to disappear, the Earth would revert to uniform motion, and fly off on a straight trajectory.

Newton's so-called second law of motion states that the acceleration of a body is in direct proportion to the force acting upon it. The constant ratio of force to acceleration is called the *inertial mass* of the body, so that

$$\text{force} = (\text{inertial mass}) \times (\text{acceleration}) \qquad (1.1)$$

Thus when the force vanishes, so does the acceleration, and the body continues with uniform velocity. Equation 1.1 is unchanged by adding a *constant* velocity to the body, for it is only the *change* of velocity – the acceleration – which enters into the second law. Equation 1.1 also expresses our experience that more massive objects are harder to accelerate with the same applied force (it is harder to push a motor-car than a bicycle). Of course, the equation cannot be solved until the nature of the force is specified, because this force may vary with position or time.

If we read Newton's second law from right to left it states that a uniformly moving system exerts no forces on its component parts. Hence the behaviour of these parts (including perhaps human beings), cannot be changed by an overall uniform motion. Thus one state of uniform motion cannot be distinguished from another by any *mechanical* means.

The reason why these principles of motion are not immediately obvious is due to the fact that in everyday experience on Earth they appear at first sight to be contradicted. For instance, in order to make a motor-car travel along a horizontal road at a uniform velocity of 50 kilometres per hour we certainly require a motive power – in this case a petrol engine. Is this experience in contradiction to Newton's laws which assert that such uniform motion will be naturally perpetuated, and only an acceleration of the car requires a special motive power? The answer lies in the fact that a motor-car requires motive power even to maintain a uniform velocity because it is necessary to overcome the forces of friction and air resistance which are always present in practice, and which in accordance with the principles outlined above will, in the absence of this motive power, tend to decelerate the motor-car until it comes to rest (with respect to the road!). In the case of the movement of the planets around the sun, such frictional forces are negligible and need not be considered. This is because the Earth moves through a nearly perfect vacuum, and not through a material medium which might interfere with its motion. The same applies to a space capsule, which, after receiving an initial acceleration from rocket motors to send it on its way, continues on its journey through space, unaffected by frictional deceleration, and requiring no further motive power. *Space itself* exerts no force on a moving body.

The fact that in the absence of motive power, terrestrial mechanisms tend to run down through dissipative frictional forces has led to the misconception that there is a natural state of rest in the universe which is the state reached when a 'moving' system exhausts its energy supply and 'stops'. Even fairly sophisticated science fiction stories have done much to promulgate this misconception, by insisting on equipping spacecraft with 'drives' and rockets which burn continuously to maintain the uniform motion of the craft through space. In one famous television series, the fate that befell a spaceship whose energy source failed was to stop 'dead in space'. There is no doubt that this sort of nonsense does a great disservice for the education of the public. At a time when Newton's mechanics has itself been superseded by Einstein's theory of relativity for three-quarters of a century, it seems incredible that some fiction writers have not yet caught up with Newton's principles which are now about 300 years old.

How do Newton's laws of motion, with their spectacular success in correctly describing the paths of the planets round the sun, support his model of space and time? The mechanical properties of bodies cannot provide a means of either locating the position of the body in space, or its velocity through space. This is clearly powerful support for the relationist theory, which would require all reference of motion to space itself to be meaningless. Nevertheless, Newtonian mechanics does apparently provide for some motions through space to be detectable, namely accelerated motion. Acceleration leads to well-known forces, sometimes called inertial forces; for example the pressure of one's feet downwards when a lift accelerates briefly upwards, or the centrifugal force of a turning roundabout (merry-go-round) trying to pull one away from the centre. It is not necessary to inspect material bodies in the surrounding world to deduce that the lift or roundabout is accelerating.

What is the origin of these inertial forces? Newton attributed them to the space in which the acceleration was taking place. If this is correct, then even if all the contents of the universe are removed except for the roundabout, the centrifugal forces would still appear when the roundabout is rotated relative to the surrounding space. The existence of inertial forces could therefore be taken as a refutation of the relationist position and the establishment of the physical reality of space.

1.4 Mach and the relationist view

Whilst the inability of Newtonian mechanics to provide a means of locating the position of a body in space, and the measurement of its velocity through space, support the relationist picture, the presence of inertial effects seems to support the Newtonian model of space as a substance which can act on bodies in at least certain states of motion.

However, a closer inspection reveals a somewhat ambiguous picture. After all, to say that in a universe devoid of all matter except a roundabout centrifugal forces remain, is a totally unfalsifiable proposition. There is only one universe, and we cannot remove all the matter from it. Thus, we can equally well replace the notion of acceleration (i.e. rotation) relative to Newton's space with that of acceleration relative to the rest of the matter in the universe. This was the standpoint taken by the nineteenth-century philosopher and physicist Ernst Mach (Austrian, 1838–1916). Mach attempted to support his thesis by appealing to a well-known observational fact. If a pendulum suspended freely from a pivot is set into oscillation at a fixed location on the surface of the Earth it will not continue to swing back and forth in the same plane indefinitely. Instead the direction of the oscillations slowly rotates (anticlockwise in the Northern hemisphere) so that it completes one revolution in one day. Anyone who is sceptical of this may see such a pendulum at the Science Museum in London. The explanation of this phenomenon lies in the rotation of the Earth on its axis, but the significant fact is that the pendulum's plane of oscillation remains fixed *relative to the distant stars*. The pendulum, being freely pivoted, is unaccelerated whilst the Earth spins beneath it.

We thus have the observational fact that a mechanical system which experiences no inertial forces also happens to be unaccelerated relative to the distant stars (galaxies really, as the stars are known to be in slow rotation about the galaxy). In Newton's model of space this fact is treated simply as a coincidence, but for Mach it assumed the greatest significance. Not only did it suggest to him that statements of acceleration relative to space could be replaced by statements of acceleration relative to the distant stars, it also implied that local mechanical systems (e.g. the pendulum) must be influenced by the distant matter in the universe in order to 'know' what was the local non-accelerating reference frame. Thus Mach attributed the inertial forces experienced by accelerating systems to an *interaction* with the distant matter in the universe. This is a remarkable theory, implying as it does that the force pushing back against you when you try to push a motor-car is caused by the action of galaxies thousands of millions of light years away! According to the Machian view, if these galaxies were removed the inertial forces would disappear. A roundabout in an otherwise empty universe would not hurl its occupants off no matter how fast it was spun around. Indeed, the whole concept of rotation in such a world is assumed to be meaningless.

Mach never elevated his conjecture to the status of a physical theory (he didn't even specify the nature of the interaction) though many people have since attempted to do so. In chapter 4 it will be seen how the natural interaction to invoke for the operation of Mach's principle is gravitation. The currently accepted theory of gravitation does not seem to incorporate Mach's ideas in a very convincing way. Even though this failure lends some support to the meaningfulness of *acceleration* relative to empty space, there does not seem to be any justification from

Fig. 1.7. The origin of inertia? The man on the rotating roundabout in (*a*) feels a force trying to drag him off. What causes this force? Ernst Mach noticed that the man also sees the stars spinning round; the force stops when the stars stop. So do the stars cause the force? If so, then in an empty universe (*b*) the force would disappear. The presence of such forces when a body accelerates (e.g. rotates) endows the body with inertia, or inertial mass. This idea is still just a speculation.

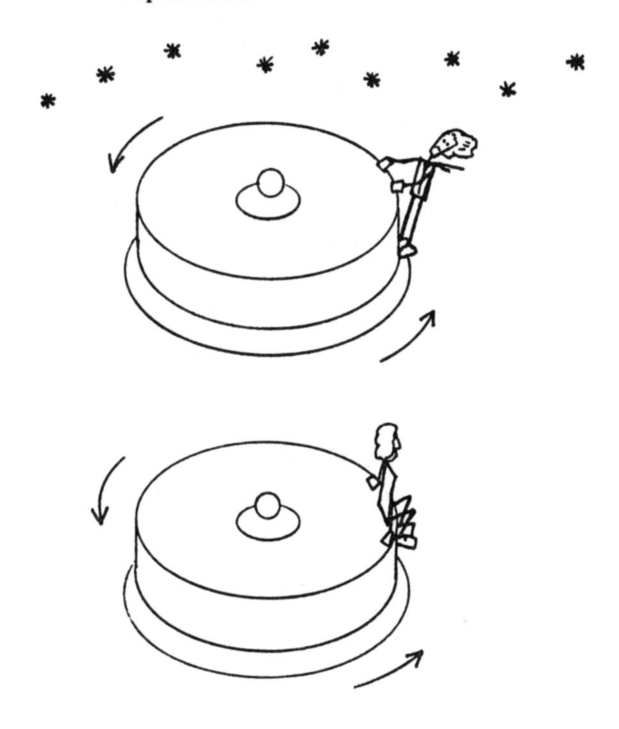

the laws of *mechanics* for Newton's full model of space in which objects also have a definite location in space, and velocity through space. A more economical model, a neo-Newtonian picture, is a space in which certain motions (accelerated) are distinguished (by the presence of inertial forces) while other motions (uniform) form a privileged class (no inertial forces). Members of this class are referred to as *inertial* motions and their associated reference frames as *inertial reference frames.* Rather than regarding space as a substance, it is more accurately pictured as an agency for differentiating between these different types of reference frame.

In the neo-Newtonian picture of space and time it is meaningful to talk about the separation of two events in time, even though they may occur at different places. But it is no longer meaningful to talk about the separation of two events in space unless they occur *simultaneously.*

To understand this remark, think of an *event* as being something occurring at a definite moment in time, and a definite location in space; for example, the chiming of a clock. The two events characterised by the chiming of the clock at one and at two o'clock are, in Newton's model, quite unambiguously separated by a time interval of one hour whether we are sitting in front of the clock on both occasions, or sitting on the surface of the sun (rather a hot seat perhaps!). However, asking for the spatial separation of the two events, the answer is not at all clear. For if we sit in front of the clock, we should not regard it as having moved anywhere in the hour that elapses between the two sets of chimes, and we should be inclined to say that the events occurred at different times but at the *same* place (e.g. in my living-room). But it is also true that during the hour which elapses the Earth has moved about 30 000 kilometres in its journey around the sun, so that an observer fixed to the sun would regard these same two events as occurring at a one hour separation in time and a 30 000-kilometre separation in space. Both agree on the time difference, but do not agree on the spatial separation.

If the physical world were described by Newton's laws of mechanics alone, one would be compelled to reject Newtonian space in favour of neo-Newtonian space. However, it is possible that other physical phenomena are capable of discerning the structure of space in a way which is independent of the motion of material objects. These other phenomena when interacting with matter might then be used to determine, for example, the velocity of the Earth through Newtonian space. To appreciate the possibility of the existence of such phenomena, it is necessary to make a short digression into the theories of gravitation and electromagnetism.

1.5 Newton's theory of universal gravitation

One of the most spectacular successes of Newton's theory of mechanics was its ability to correctly describe the motion of the planets in the solar system under the action of the force of *gravity*. From his observations of falling bodies (and, so the story goes, a falling apple) Newton proposed a theory of universal gravitation. According to this theory all material bodies in the universe attract one another with a gravitational force. Some of the features of this gravitational force may be deduced from common experience. A plumb line pulled towards the surface of the Earth by the Earth's gravitational force lies vertical, indicating that the force between spherically shaped material bodies lies along the line joining their centres. The force of gravity between bodies may be attributed to a gravitational 'charge', in the same way as electrical forces exist between electrically charged objects. Galileo Galilei (Italian, 1564–1642) discovered the important observational fact, to be discussed in detail in chapter 3, that bodies dropped together near the surface of the Earth reach the ground together, i.e. undergo equal accelerations. A glance at Newton's second law, equation 1.1, shows that equal accelerations under gravity requires the gravitational force to be proportional to the inertial mass of the body. To put it simply, a more massive object is harder to accelerate downwards than a less massive object, but the gravitational force which acts is greater in direct proportion to the mass, thereby compensating exactly. We may express this fact by saying that the gravitational charge is proportional to the inertial mass, a fact which will prove to be of the greatest significance.

Finally, an elementary knowledge of the behaviour of the

planets in the sky indicates that the more distant planets take longer to revolve once around the sun than do those near to the sun. Consequently the gravitational force must diminish with distance.

The law of gravitation which Newton proposed was that the magnitude of the force F between two (compact) bodies of masses m_1 and m_2 a distance r apart is given by the equation

$$F = \frac{Gm_1m_2}{r^2} \tag{1.2}$$

where G is a constant with the same value for all bodies in the universe. It is known as Newton's constant of gravitation and is the proportionality constant (mentioned above) required to convert units of mass into units of gravitational charge.

A fundamental and far-reaching assumption which Newton made about the force (1.2) was that it acts *instantaneously* across the empty space between the two bodies. It is therefore a theory of instantaneous action-at-a-distance. Instantaneousness is a well-defined concept according to Newton's model of time, because there is unambiguous meaning to the notion of the 'same moment', even at points separated by the distance r. The time measured in the vicinity of both bodies is the same universal time.

By combining together his law of gravitation (1.2) with his fundamental law of motion (1.1), Newton was able to predict that the path of a planet around the sun would be an ellipse. This is correct, and represented a spectacular success for Newtonian mechanics, as well as for philosophy, for it showed that even the 'heavenly bodies', as they were then regarded, obeyed 'down-to-Earth' physical laws which could be established in a terrestrial laboratory. This lesson of history has been repeated again and again as new laws of nature are discovered on Earth and found to be true even in the most distant regions of the universe that we can see.

1.6 Maxwell's electromagnetism and the ether

Despite the enormous successes of action-at-a-distance in explaining the motion of the planets around the sun under the action of an instantaneous gravitational force, Newton's theory could *not* explain correctly the closely analogous situation of the motion of electrically charged particles acting on each other across empty space with electric and magnetic forces. For just as the Earth is constrained to deviate from straight line motion by the gravitating influence of the sun, so a charged particle such as an electron will also be forced along a curved trajectory under the action of electric and magnetic forces. Indeed, the modern picture of the atom is in many respects similar to that of the solar system, with the heavy, positively charged atomic nucleus at the centre of the atom (representing the sun) and the lighter mobile electrons revolving around the nucleus at great speed.

Nevertheless, there are three important differences between gravitational and electromagnetic forces. Firstly, only certain particles carry electric charges, whereas all forms of matter and energy possess gravitational charge. Secondly, electric forces may be both attractive and repulsive so that we label one set of charged particles which have the same type of charge *positive*, and the other set *negative*. Positive attracts negative, but both positive and negative charges repel their own kind. On the other hand, gravitating objects always attract each other. The reason that like charges repel in the former case can be traced to the fact that the electromagnetic force is a *vector* force, i.e. its direction may also vary as well as its strength (this is why we must introduce the idea of magnetic as well as electric forces) whereas Newtonian gravity *always attracts* along the direction of the line between the particles. The final difference between these two fundamental forces of nature concerns their relative strengths. To say that the electromagnetic forces in an atom are much stronger than the corresponding gravitational forces is a sheer understatement. They are stronger by 10^{39} (recall that this is one followed by 39 zeros). For this reason almost all the phenomena of everyday life are dominated by electromagnetism. However, owing to the existence of opposite electric charges, large accumulations of matter such as the Earth are more or less electrically neutral. In contrast, the gravitational charge of the constituent atoms, whilst exceedingly feeble, is always cumulative. By the time one has reached the size of the Earth – with its 10^{51} atoms –

gravitation overwhelms the intrinsically stronger electromagnetic effects from sheer weight of numbers.

The inability of the early physicists to formulate an action-at-a-distance theory of electromagnetism is related to these differences, and also intimately to the subject of *time asymmetry*, as will be seen in chapter 6. In recent years this deficiency has been redressed, so that it is possible to recast the theory of electromagnetism, and the related subject of electrodynamics (the motion and action of charged particles) in a type of action-at-a-distance language.

The great breakthrough in the understanding of electromagnetic forces came with the brilliant mathematical insight of the nineteenth-century British physicist James Clerk Maxwell (1831–1879). Maxwell combined together the results of experimental work already carried out by Hans Oersted (Danish, 1777–1851), Michael Faraday (English, 1791–1867), Heinrich Lenz (Russian, 1804–1865) and others into a unified set of mathematical equations which beautifully and correctly describe the interplay between the motion of charged particles and the behaviour of electromagnetic forces. Central to the whole theory of Maxwell was the introduction of a profound and far-reaching new concept in physics – the *field*. By casting the laws of electromagnetism in the language of fields Maxwell at once swept aside the difficulties associated with Newtonian action-at-a-distance and opened an entire new chapter in the history of physical science.

A field is in some senses more abstract than a particle. Maxwell conceived of all charged particles being surrounded by their own electromagnetic fields rather like an invisible halo, the presence of which was quite undetectable except when other charged particles were placed in it. The action of the field was then to exert a force upon the charges. Nineteenth-century physicists liked to describe the field in close analogy to the motion of fluids. Consequently, there emerged the use of words like magnetic *flux* and lines of force (similar to fluid streamlines) – words which are still with us today. But a fluid description implies a *medium* of some sort to transmit the action of one charge upon another. Belief in this fluid was so well established in the nineteenth century that it was given a name – the *ether*. This medium was supposed to fill all of empty space although it was meant to be quite invisible. Electromagnetic fields were to be thought of as *stresses* in the ether. Even more exciting, was a new possibility. In a more familiar medium such as air, a disturbance may set up vibrations of stress waves (ordinary sound waves) travelling outwards through the air from the region of disturbance, at a fixed speed depending on the elastic properties of the air. So too would a disturbance of charged particles set up stress waves in the hypothetical ether. Moreover, the speed of these waves could easily be calculated from the electric and magnetic properties of the medium, in this case 'empty' space. At the time that Maxwell predicted the existence of these electromagnetic waves the values of the appropriate quantities were known and the speed of the ether waves turned out to be very fast indeed – about 300 000 kilometres per second. We now believe this to be the fastest speed in the universe (at least for ordinary objects). But in Maxwell's day it had another important significance. It was just about the speed that Olaus Romer (Danish astronomer, 1644–1710) had determined for the propagation of light signals by observations on the orbits of the moons of Jupiter! Physics had made a great leap forward. Light apparently consisted of electromagnetic waves, generated by the motion of charged particles and travelling across space in the form of a vibration of the ether.

Nor was this all. Just as sound waves may vary in pitch (frequency) so too can electromagnetic waves. In fact, small variations in the frequency of light waves lead to differences in the colour quality of the light. But what about large changes? Light itself is produced by the violent high frequency disturbances which take place inside atoms when they become excited, as for example under the action of intense heat. But in the laboratory much less violent disturbances may be generated with electrical mechanisms which move charged particles around at relatively low frequencies. Can these waves be detected? Indeed they can; they are none other than radio waves, searched for and eventually produced by Heinrich Hertz (German, 1857–1894) some 20 years after Maxwell had predicted their existence. Today the entire spectrum of frequencies of these electromagnetic waves is familiar to the physicist. As well as radio and microwaves, there is heat radiation (infra-red), ultra-violet, X-rays and γ-rays. All are well understood in the context of Maxwell's theory which is now over 100 years old.

With the arrival of electromagnetic waves, physics had at its disposal a decisive method of checking on Newton's model of space and time. The crisis which accompanied this attempted verification at the turn of the century, and the brilliant mathematical and physical edifice to which this crisis gave birth, marked a turning point in the history of physics and human thought which ranks with the Newtonian revolution of two centuries before. This new revolution was the theory of relativity.

2 The relativity revolution

2.1 Space and time in crisis

The position or velocity of a body in Newton's absolute space cannot be revealed by any mechanical experiment. However, with the appearance of Maxwell's electromagnetic theory, there arose the possibility of using optics – the motion of light signals – to measure the velocity of bodies through space. The success of such a venture depended crucially on Maxwell's picture of an ether, a kind of fluid filling all of space (presumably at rest). The motion of a body through space could be deduced from its motion in the ubiquitous ether.

Specifically, it was expected that the velocity of the Earth through the ether could be determined on the basis of the following reasoning. As the Earth orbits the sun, it moves with varying velocity through the ether. From the standpoint (reference frame) of an observer on the surface of the Earth, the ether flows past in a 'slipstream' or current; a very evanescent stream to be sure, for Newtonian mechanics forbids it to exert any force or drag on the moving Earth, or it would slow up and drop into the sun. Nevertheless, in the nineteenth century the ether stream was considered very real. The challenge was to measure its rate of flow. Maxwell's theory predicts that light travels through the ether with a fixed speed depending only on the 'elasticity' of this medium. It follows that the speed of light as measured by an Earthbound observer will vary depending on the direction in which the light is travelling. For example, light which moves 'downstream' in the ether is swept along at a greater speed than oppositely directed light moving 'upstream' against the current.

Efforts to determine the speed of the ether stream were pursued with great ingenuity. In 1887 the most famous of these experiments was carried out by the two American physicists Albert Michleson (1852–1931) and Edward Morley (1838–1923). The principle of their experiment is best explained in analogy to an ordinary river. If a swimmer crosses a river from bank to bank and back again, he will always arrive back before a colleague who, swimming with equal speed relative to the water, travels the same distance upstream and then downstream (see fig. 2.1). The reason for this is a simple one. The latter swimmer has to contend with the current, which helps him on the downstream leg, but hinders him on the upstream leg. Because the upstream journey is slower it takes longer, so the hindrance lasts slightly

Fig. 2.1. Michleson–Morley light beam race. Peter and Paul both swim four metres per second in still water. They leave the bankside point *A* together, Peter swimming downstream to *B* and back, Paul swimming the same distance cross-stream to *C* and back. Paul always wins the race. The reason is simple. The river flows at, say two metres per second. Paul arrives at *C* after 10 seconds, but Peter, helped by the current and travelling at six metres per second relative to the bank, arrives at *B* after only $6\frac{2}{3}$ seconds. On the return journey Peter more than loses his lead. Paul takes the same 10 seconds to get back, making a round trip time of 20 seconds in all, but Peter, slogging against the current, is slowed down to two metres per second relative to the bank and takes a full 20 seconds just to get back to *A*. His round trip time is thus $26\frac{2}{3}$ seconds – $6\frac{2}{3}$ seconds behind Paul.

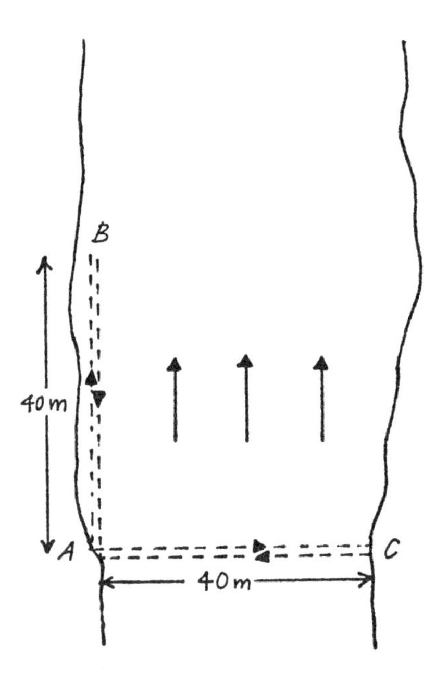

longer than the help. The result is that he arrives back later than his colleague, who is neither hindered nor helped in his cross-stream exertion.

Michleson and Morley used light beams as the 'swimmers' in the ether stream, sending them out and back in perpendicular directions. A careful measurement of the difference in their travel times can be made by superimposing the two returned beams. The result of this experiment was certainly perplexing. With the benefit of hindsight, it clearly represents the decisive blow of experimental physics which shattered the (then) 200-year-old edifice of Newtonian space and time. Although the minimum effect expected from the Earth's orbital motion was well within the capability of Michleson and Morley's apparatus, they quite simply found nothing at all. The ether had stopped flowing – it has never started again since.

The absence of an ether stream faced physics with a fundamental and disturbing inconsistency. Efforts to patch up the ether theory paled into mediocrity when challenged by a new and powerful mind. Albert Einstein (German, 1879–1955), one of the world's greatest physicists, exploded the whole conceptual framework in which the ether stream experiments had been conducted. What Einstein proposed amounted to nothing less than a complete abandonment of the familiar Newtonian picture of space and time which had endured so successfully for so long. In its place was presented a theory of space and time which leads to strange and unfamiliar consequences. The basic form of this new theory was published in 1905. Einstein called it the special theory of relativity. It opened a new chapter of science.

At the heart of the special theory of relativity is a denial of the reality of Newtonian space. The ether cannot be detected because it is not there. The whole notion of an absolute frame of rest against which the velocity of an object through empty space could be measured is a fiction. Uniform motion is only to be reckoned *relative* to some other material system. Not only mechanics, but also any physical experiment whatever, is totally unable to detect the absolute velocity of a system through space. The entire concept of uniform motion through a fixed space is without meaning. Thus did the ether, along with alchemy and phlogiston, pass into the realm of historical curiosities. In its place Einstein proposed a curious new principle.

The new principle of special relativity at first seems innocuous, and then utterly baffling. It states that the speed of light is everywhere constant. By this is meant that it has the *same* speed, whether measured on Earth, or in a rapidly moving rocket, whether the source is at rest or speeding towards or away from the observer. Even two observers side by side, passing each other at great speed, measure the *same* speed for the *same* light beam.

Contrast this situation with the previous assumption that light travels at a fixed speed through space. In that case a man travelling in a rocket very fast (see fig. 2.2) towards a beam of light would obviously meet it head on at a greater speed than a colleague in a similar rocket travelling in the opposite direction, after whom the light beam would have to chase and overtake. Such a situation is certainly true of sound waves and most people would not question its validity with light waves also. Einstein said this was incorrect. Both rockets encounter the light moving with the same speed! It follows that however energetically a rocket chases after a light beam, the light still runs away from the rocket at the same speed which is, moreover, no faster than the speed with which it runs away from the rocket moving in the opposite direction.

Clearly this principle of the constancy of the speed of light explains the failure of Michleson and Morley to detect any

Fig. 2.2. Light speed is the same for everyone. The astronaut chases the light pulse. It gains on him at 2.998×10^8 metres every second. The cosmonaut flees the *same* pulse. It leaves him at 2.998×10^8 metres per second. The astronaut accelerates ('I'll overtake it yet'). The light still recedes at 2.998×10^8 metres per second. He is no better off than the cosmonaut driving hard in the opposite direction!

difference in the time required for two light beams to travel on a round trip through the 'ether'. Both beams travel equally fast irrespective of the direction of the Earth's motion. However, the principle is clearly nonsensical unless the whole idea of fixed space and time is discarded. Something very peculiar must occur if a rocket cannot gain one little bit on a light beam, however powerful the engines it possesses.

2.2 The overthrow of Newtonian time

Just how peculiarly things do behave when the constancy of the speed of light is assumed is often illustrated by considering the experience of an individual travelling at great speed in a railway carriage (no doubt the humbler choice of transport traditionally adopted in discussions of this sort reflects the fact that the most rapid vehicle generally available in 1905 was the railway train!).

To obtain a noticeable effect, the train must be travelling very fast indeed – at a sizeable fraction of the speed of light, relative, of course, to the railway track. The fact that such rapid transportation is not in practice available in the twentieth century (even the Earth moves around the sun at a minute fraction of the speed of light) only serves to explain why nothing unusual is noticed when travelling about in everyday life, and why the theory of relativity took so long to be invented. Nevertheless it shall be assumed that we are dealing with a supertrain.

Riding inside a railway carriage is an observer called A (see

Fig. 2.3. Simultaneity is relative. As far as A is concerned the two light pulses hit the carriage ends at the same moment, because they travel equally fast *in the carriage*. B thinks they travel equally fast *along the track*. In the brief flight time (perhaps one tenth of a microsecond!) the supertrain has shot forward to the position marked by the broken lines. The left-hand pulse appears to B to hit the rear end of the carriage well before the right-hand pulse has caught up with the front end.

fig. 2.3) equipped with a lamp, which he fixes at the exact centre of the carriage. On the embankment waiting for the supertrain to rush by is a colleague called *B*, who can also see the lamp and the inside of the carriage. An experiment is arranged between *A* and *B*. The supertrain will move at a constant speed past *B*. At some convenient moment when the carriage passes by, the lamp in the centre flashes briefly, sending a short-duration light pulse in both directions down the carriage. When these pulses reach the carriage ends the event is recorded by both *A* and *B*.

The outcome of this imaginary experiment is rather odd. *A*, who is travelling in the train, sees both pulses leave the centre and pass down the inside of the carriage, reaching both ends at the same moment. This happens because the speed of light is the same for both pulses and the distances over which they travel are equal. The experience of *B*, standing beside the railway track, is very different. He also observes the pulses travel in opposite directions at the same speed, but this speed is now reckoned by him in his own reference frame, fixed to the railway track. Consequently, the left-hand pulse (see fig. 2.3) reaches the left-hand end of the carriage *before* the right-hand pulse reaches its end. The reason for this is that relative to *B* (though not *A*) both the train and light source are in motion, so that during the time which the light takes to travel down the carriage the train moves forward somewhat. The left-hand pulse thus has less distance to travel than the right-hand pulse so, their speeds being equal, it arrives first.

A and *B* experience different versions of the same events. Who is right? Do the light pulses really arrive at their respective ends simultaneously, or does the left-hand one arrive first? According to the theory of relativity, *both* are right. We cannot say '*A* is moving, so he is mistaken' because *A*'s uniform motion is physically irrelevant. It is only motion relative to *B*. One might just as well say that the train is at rest and the Earth is moving past in the opposite direction. Indeed, this may well be more persuasive as the Earth certainly is moving round the sun. The special theory of relativity demands that there are no privileged reference frames, nobody has the special status to be 'right' and everybody else who is moving differently 'wrong'. The inescapable conclusion is that certain occurrences which are traditionally considered to happen objectively in a certain way are not objective at all, but merely *relative* to a particular state of motion. In particular, the *simultaneity* of two separated events is not an absolute property possessed by the events themselves, but only a consequence of the way the events are observed. What constitutes the *same* moment at the two carriage ends relative to *A* is not the same moment as observed by *B*.

This new view of time is rather bizarre when first encountered. Before the theory of relativity everyone assumed that the time being used by a railway passenger, a trackside spectator and a man on Mars were all the same. Newtonian time is absolute and universal; unchanged by one's state of motion and fixed throughout the universe. This view of time as a fixed background or framework against which to measure events is now known to be wrong. A universal 'same' moment does not exist.

Further disturbing consequences of the relativity principle follow if we allow for a second train, carrying a third observer called *C*, to pass by on a parallel track. If *C* is travelling faster than *A*, he will overtake *A*'s train so that, *relative to C*, *A* will be travelling in the reverse direction, i.e. from right to left in fig. 2.3. It follows from an identical argument to that used above that *C* will observe the two light pulses on *A*'s train to arrive at the carriage ends at different times. Only in this case, because the motion is right to left, it is the *right-hand* pulse which arrives at its destination first. So the *order* of events is reversed from *B*'s experience. *B* sees the left-hand pulse arrive before the right-hand pulse and *C* (equally correctly) sees it the other way around. Thus the theory of relativity demolishes the before–after relation between spatially separated events. It does not, however, destroy this relation for events that occur at the *same* place or, as it turns out, for those events that can be joined by a signal travelling at, or less than, the speed of light. In fact, an observer cannot change his state of motion so that he sees time running backwards in another reference frame. What his state of motion does affect though is the *rate* at which clocks are observed to run. Although the time order of two events at the same place (e.g. ticks of a clock) is invariant, the *duration* between them is not. The reason for this may be easily understood using elementary algebra, but the reader who is unhappy about sums may wish to skip the demonstration and pass directly to the discussion of the effect which follows after equation 2.6.

Consider two inertial (recall this means uniformly moving) observers A and B in relative motion with velocity v in the x direction (see fig. 2.4). Each observer can be regarded as fixed to an imaginary rigid reference frame against which distances may be measured. Observer A measures distances x from the origin of his frame, labelled S, while B measures distances marked x' from the origin of his frame, labelled S'. Naturally x and x' will be different, except at the moment when S and S' coincide, which we shall take, without loss of generality, to be at time $t = 0$. At a later time t we might expect x and x' to be related by

$$x = x' + vt \tag{2.1}$$

because we must add to x' the distance vt which the frame S' has travelled relative to S in the duration t. Equation 2.1 is the correct relation according to Newtonian physics but it is necessary to take account of the fact that the time t as measured by A might not coincide with the time, say t', as measured by B, because of the relativity of simultaneity. Moreover, equation 2.1 could not possibly be consistent with the speed of light remaining the same for *both* A and B. The simplest generalisation of (2.1) which allows for these two effects is

Fig. 2.4.

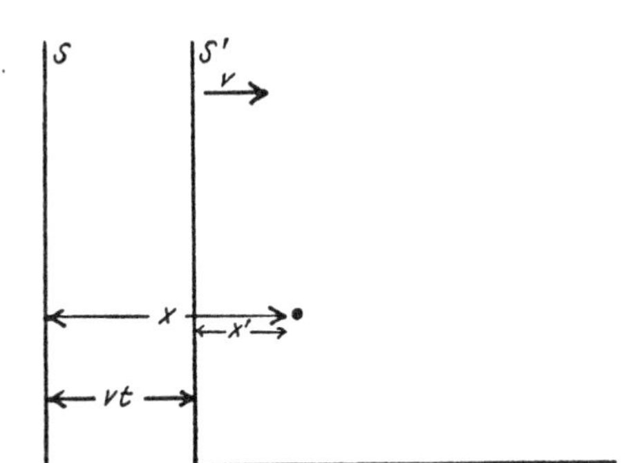

$$x = \gamma(x' + vt') \qquad (2.2)$$

where γ is a factor which approaches one when v is very small, because we know that (2.1) is correct in the limit of small velocities v, since Newtonian physics is such a good description of 'everyday' physical systems.

Now because of the relativity of all inertial motion, the relation (2.2) must be symmetric between S and S', for we could equally well regard S' as fixed and S moving to the left with velocity v. From the standpoint of B therefore, we require

$$x' = \gamma(x - vt) \qquad (2.3)$$

where the minus sign in front of v denotes motion to the left.

It is now possible to see how equations 2.2 and 2.3 are consistent with the requirement that the speed of light, called c, is the same for both A and B. This requirement may be expressed by saying that $x = ct$ implies $x' = ct'$. Substituting these values for x and x' into (2.2) and (2.3) gives

$$ct = \gamma t'(c+v), \qquad ct' = \gamma t(c-v)$$

from which we may eliminate t and t' to give

$$\gamma = \frac{1}{\sqrt{(1-v^2/c^2)}} \qquad (2.4)$$

This factor does indeed have the property that it approaches one as v approaches zero. Notice that $x = ct$ in the *Newtonian* equation 2.1 implies

$$x' = (c-v)t$$

with the consequence that B measures the speed of light as $c-v$ rather than c in Newton's theory.

To understand the significance of the factor γ for the measurement of time intervals by A and B, we may first eliminate x' from (2.2) and (2.3) to give

$$t' = \gamma\left(t - \frac{vx}{c^2}\right) \qquad (2.5)$$

which relates the time t' as measured by B to the time t measured by A at the point x. For events fixed at definite points in the frame S, x does not change. This means that the time interval between two events 1 and 2, say $t'_2 - t'_1$, as measured by B is related to the corresponding interval $t_2 - t_1$, measured by A by

$$t'_2 - t'_1 = \gamma(t_2 - t_1)$$

which we may abbreviate to

$$\Delta t' = \gamma \Delta t = \frac{\Delta t}{\sqrt{(1 - v^2/c^2)}} \qquad (2.6)$$

$\Delta t'$ and Δt denoting the respective intervals.

It is clear from equation 2.6 that the intervals Δt and $\Delta t'$ are not the same unless $v = 0$. That is to say, the relative motion between the observers A and B causes them to measure *differently* the intervals of time between two events. The effect, which is contained in the factor γ, is clearly very small for 'everyday' velocities v, which are very much less than the speed of light, c, because the number v/c is then tiny. For example, even a rocket travelling at 50 000 kilometres per hour corresponds to v/c equal to only 0.00005 and γ equal to 1.000000001. That is to say, an observer in a rocket travelling away from Earth at 50 000 kilometres per hour would observe intervals of time on the Earth to be fractionally extended by a mere ten-millionth of 1 %.

Another way of expressing this phenomenon is by the rate of time as measured by clocks. A clock carried in a fast rocket would run slow relative to a similar clock left on Earth. It is important to realise that the effect here is a property of space and time, and has nothing to do with the mechanism of the clock itself. An observer travelling with the clock in the rocket would not regard his clock as behaving unusually. Indeed the measurement of time intervals, whether by the clock, his own brain, or any other process, would be quite consistent and unremarkable, as we know it must be in view of the fact that the Earth itself is actually moving at very great speed relative to distant galaxies, but without producing any bizarre temporal effects. The theory of relativity which predicts the slowing of clock rates is founded upon the relativity of uniform motion, so that no *internal* observation of clocks or anything else is permitted to indicate whether a system is moving or not – remember that there is no absolute uniform motion. The clock effect, usually referred to as the time dilation effect, is only manifest when observers may inspect *other* systems relative to which they are in motion. Thus the observer in the rocket sees the Earthbound clock to be running slow, but not his own, while, because of the symmetry inherent in relative motion, an observer on Earth would likewise see the clock in the rocket running slow relative to his own well-behaved clock. It will be seen on inspecting equation 2.6 that when v approaches c, γ becomes indefinitely large, so that the interval of time Δt appears to the moving observer to be indefinitely extended.

In the limit of $v = c$, γ is infinite, which is to say that an observer who travelled at the speed of light would see all clocks permanently immobilised – the passage of time would be stopped altogether. For this reason it is sometimes said that a light beam experiences no time at all to travel any distance, however large.

2.3 The twins 'paradox'

The apparently paradoxical conclusion that both clocks are running slow relative to each other sometimes causes a certain amount of confusion to the unaccustomed reader. It must not be imagined that the time dilation is an *illusion*, caused by the propagation of light signals or whatever. It is not that each observer merely *sees* the other clock running slow, it actually *is* running slow – a real physical effect. A dramatic way of illustrating this is to arrange for two identical twins to participate in an experiment in which one twin leaves Earth in a fast rocket which travels at near the speed of light to the nearest star (in the constellation of Centaurus) and back again. The time for this round trip as measured by the twin on Earth is several years, but for the twin in the rocket this travel time may be made as short as he pleases by closely approaching the speed of light. Thus the travelling twin will return to Earth several years younger than his stay-at-home sibling, having only experienced a short travel duration as against several years duration on Earth.

It is fascinating that the special theory of relativity opens up the possibility of time travel. Indeed, with sufficient resources to achieve near-to-light rocket speeds, anyone may travel indefinitely far into the future in this way. It is an amusing fact that people who live at high altitudes travel (over their whole lifetime) about 10^{-12} seconds into their sea-level counterpart's future, solely on the basis of the increased speed at high altitudes from the Earth's rotation. However, it is not possible to use the time dilation effect to travel back again into the past. Only one-way trips are allowed.

At this stage the reader may well be puzzled as to how the space traveller can both regard the clock on Earth as running *slow*, and yet return to Earth to find his twin *older* (rather than younger) than he. This apparent contradiction has long caused puzzlement to students of the special theory of relativity, and has earned the name 'the twins paradox'. It is not really a paradox at all, as is revealed by a careful consideration of what the two observers actually see. Two observers will literally see each other by sending light signals to and fro. If the observers are separated by a large distance, there will be a noticeable delay in the reception of these signals due to the finite time which the light takes to travel across the intervening gap. To fix ideas, note that light takes about a second to travel from the moon to the Earth and about eight and a half minutes from the sun. It takes just over four years to reach us from the nearest star. This star is about 40 million million kilometres away, but it is more convenient to describe this as four light years. This delay means that when we observe the nearest star from Earth we do not see it as it is now, but as it was four years ago. Likewise hypothetical inhabitants of a planet orbiting this star would at present be seeing the sun as it looked four years ago. (Notice that a common present is being assumed here for the Earth and the nearest star, which is known to be a dubious concept according to the relativity of simultaneity. However, because the relative velocity of these two objects is so low compared to that of light, the ambiguity introduced may be ignored.) The effect of the finite travel time of light is therefore to introduce a lag in synchronisation between clocks kept by distant observers. The clock *rates*, however, are unaffected, if the clocks are relatively at rest.

Consider now that the two observers are receding. With the progressively widening distance between them, the lag effect starts to grow, and the clocks gradually fall further behind each other in synchronisation. This changing synchronisation thus makes receding clocks appear to run *slow*. This new effect may be comparable with, but is quite distinct from, the time-dilation effect. Indeed, if two clocks are approaching rather than receding the new effect operates in reverse, and appears to *speed up* the clock rates. Moreover, the time dilation effect is purely a consequence of the special theory of relativity, whereas the other effect occurs for any sort of wave motion. In the case of sound waves, it is familiar as the sharp drop in tone of a train whistle, or car engine, as the vehicle approaches, passes and then recedes. It is usually called the Doppler effect (after Christian Doppler, Austrian, 1803–1853), and acts as a modification of the wave vibration rate, or frequency (it is often convenient to regard the regular vibrations of a wave as the beats of a clock). In the case of light waves the drop in frequency as a light source recedes has the effect of changing the *colour* quality of the light – a shift towards the red end of the spectrum. For this reason the Doppler effect is often called the red shift in optics, and is the method whereby astronomers can tell something about the motion of very distant heavenly bodies.

Fig. 2.5. Doppler effect. The distant clock is receding from Earth. At three o'clock it is one million kilometres away at *A*. The observer sees this event about five seconds later, because of the time taken by the light to travel back to Earth. One hour later the clock has retreated to *B* – two million kilometres away. The observer does not see the clock read four until 10 seconds afterwards, because it is now twice as far away. It appears to the observer that the clock is running slow; it has lost five seconds in one hour. If the clock were approaching it would *gain* five seconds per hour.

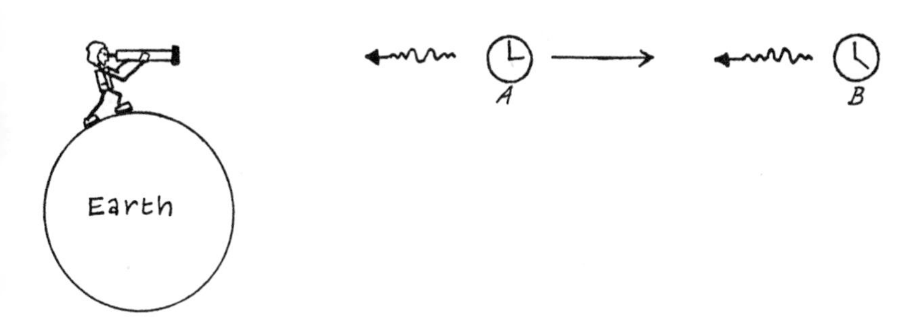

In practice, the direct observation between moving reference frames will in general involve both the Doppler and time-dilation effects together. The Doppler effect can be eliminated though, by arranging for the motion to be transverse only, i.e. across rather than along the line of sight. In most discussions of time dilation the use of the word 'see' in connection with an observation between reference frames tacitly omits contributions from the Doppler effect, and should be regarded as applying to transverse motion only. In the round-trip twins' experiment, both effects will be present.

It is instructive to analyse in detail what each twin actually *sees* during the round-trip experiment, by allowing for both Doppler and time-dilation effects. For the sake of definiteness, suppose that twin A travels to a nearby star 10 light years away at a uniform speed of $0.9c$, i.e. 90% of the speed of light. He then returns straight away to Earth at the same speed. We shall neglect the periods of acceleration and deceleration necessary to achieve such a colossal velocity, which in principle may be made to occupy an arbitrarily small interval of time. All the while, twin B remains on Earth observing A, together with his clock, as he first recedes then approaches.

Use of the time-dilation formula (2.6) shows that both twins will observe the other's clock to run 2.3 times too slow, in the absence of the Doppler effect.

Now at $0.9c$ the twin A will cover the distance of 10 light years in 11.1 Earth years as measured by B on his own clock, but because of dilation this event will be recorded as only 4.84 years on A's rocket clock. However, B will not know of A's arrival at the nearby star for another 10 years, which is the time taken for the light signals of this event to travel the 10 light years intervening distance. B, therefore, will actually *see* A's outward journey taking 21.1 years. As this is recorded as 4.84 years on A's rocket clock, B will observe throughout the outward journey A's clock running 4.36 times too slow, of which 2.3 is due to the relativistic time-dilation effect and the rest due to the Doppler effect.

To determine what A sees of B and his clock during the outward journey, note that the observations must always be perfectly symmetric between two inertial observers according to the principles of special relativity, which enable us to regard equally the situation that it is really *A* at rest, and *B* receding at 0.9*c*. Consequently, *A* will *see* events on Earth running 4.36 times too slow (again 2.3 due to relativistic time dilation). Because *A* reaches his destination after 4.84 years on his clock (note that both *A* and *B* must agree on the time on the rocket clock at which the rocket reaches the star), should he look back at Earth at the moment of arrival, he will observe events occurring only 4.84/4.36 = 1.1 years after his departure.

As regards the return journey, the two observers are approaching at 0.9*c* so although their clock rates are slowed relativistically by the same factor of 2.3 as the outward journey, the Doppler effect actually overcompensates. The whole round trip takes 22.2 years Earth time (i.e. on *B*'s clock), so that *B*, who after 21.1 years has only just seen *A*'s arrival at the nearby star, sees the entire return journey take place in just 1.1 years. Now as far as *A* is concerned the return journey must take as long as the outward journey, i.e. 4.84 years by the rocket clock. Consequently *B* will see these 4.84 years compressed into 1.1 years Earth time, so that events in the rocket during the return journey actually appear to *B* to be *speeded up* by a factor of 4.36. As regards *A*, who at the outset of the return journey is observing events on Earth only 1.1 years after his departure, he must for consistency see the remaining 21.1 years Earth time required for his journey crowded up into the 4.84 years of his return journey as measured by his own clock. This corresponds to a speeding up by a factor of 4.36 of events on Earth as observed by *A*. The speeding effect is therefore also perfectly symmetric between *A* and *B*.

We may conclude that *A* returns to Earth after 9.7 years rocket time, to find that 22.2 years Earth time has elapsed and his twin *B* is 12.5 years older than he is. At all times each observation by *A* and *B* has been quite consistent. There is no paradox, and the time dilation effect is a very real one, not merely a question of what is observed by light signals. The reason why it is *A* rather than *B* who ages less can be traced to the fact that it is *A* who changes his reference frame by accelerating to 0.9*c*, and then abruptly reversing his velocity on reaching the nearby star. Thus

Table 2.1. *The twins 'paradox' resolved. A log kept by* A *(in the rocket) and* B *(left on Earth) of the durations in years of the two journeys – out and back – as they actually* see *them on the clocks. The total round trip takes 22.2 years Earth time, 9.7 years rocket time*

	Outward journey		*Return journey*	
	Earth clock	*Rocket clock*	*Earth clock*	*Rocket clock*
A	1.1	4.84	21.1	4.84
B	21.1	4.84	1.1	4.84

although during the periods of uniform velocity the temporal effects are perfectly symmetric between *A* and *B*, the entire journey is *not* symmetric because of these periods of abrupt acceleration and deceleration experienced by *A*. Recall that acceleration is absolute in special relativity and can certainly be detected by *A* as he is hurled about in his rocket, whilst *B* experiences no such forces when remaining on Earth. The abrupt reversal of *A*'s velocity means that although the clock rates are equally slowed and speeded by a factor of 4.36 for both *A* and *B*, *A* sees the speeded-up period occur for half his journey, whereas *B* sees this period occur for only the last 1.1 out of the total 22.2 years' trip. Hence their clocks must get out of step. Readers who are bemused by the above discussion may be assisted by consulting table 2.1.

It is possible to use the round-trip experiment to illustrate another remarkable conclusion of the special theory of relativity. *A* is travelling at $0.9c$ relative to the Earth, but covers the distance from Earth to the nearby star in only 4.84 rocket years. It follows that this distance must *appear to A* to be only $0.9 \times 4.84 = 4.36$ light years, rather than 10 light years as measured by *B*. The spatial distance therefore, has shrunk by the same factor $\sqrt{(1 - v^2/c^2)}$ as the time interval has stretched. This shrinkage is known as the Lorentz–Fitzgerald contraction after Hendrik Lorentz (Dutch, 1853–1928) and George Fitzgerald (Irish, 1851–1901). It too is quite symmetric between uniformly moving observers (once again, the fact that *A* sees the distance less than

B here is due to the fact that the nearby star is assumed to be at rest relative to the Earth and not the rocket). It implies that a rapidly moving observer appears to be squashed up in the direction of motion. This squashing, like the time dilation, must not be imagined as a physical force operating on the observer, but as a property of space itself. For the moving observer feels and sees nothing unusual about his own system, but instead sees the oppositely moving surrounding world squashed up in reciprocal fashion. Note that as v approaches the speed of light, c, an object becomes completely flattened!

2.4 Faster than light?

It might be wondered what would happen to an object which was accelerated to a speed faster than light. The infinite contraction of its length and dilation of time which occur at

Fig. 2.6. Moving masses get heavier. As the speed of light is approached the whirling mass gets progressively heavier, without limit. It becomes harder and harder to speed up. All the energy in the world cannot spin it as fast as light. This is well known in laboratory experiments where the whirling masses are subatomic particles and the strong man is a machine called a cyclotron. The speeding particles really do get heavier.

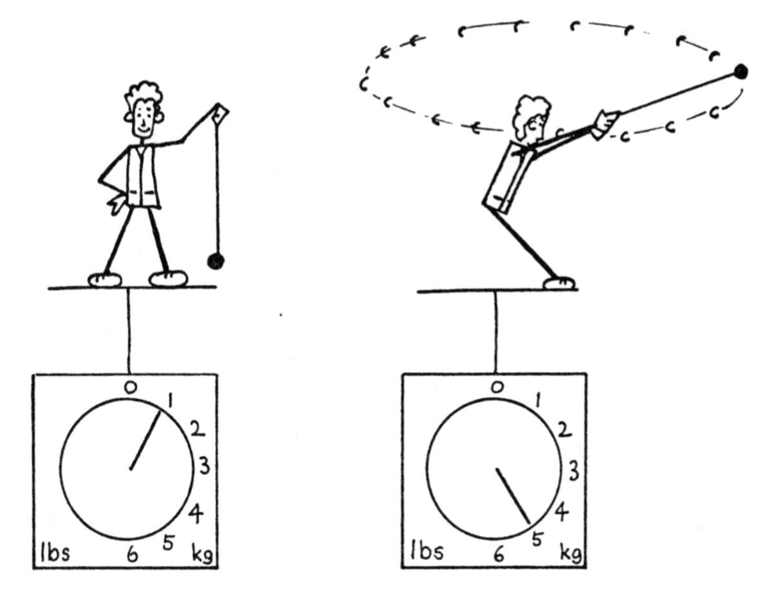

$v = c$ might be expected to place a limit on the velocity of an object to prevent it from becoming superluminal. In fact this is so. The nature of the physical barrier which is encountered at this limiting velocity is revealed if the special theory of relativity is applied to the energetics of moving bodies. It is found that the nearer the velocity approaches to that of light, the greater is the *energy* required to accelerate it faster. It would require an infinite amount of energy to reach the speed of light. This escalating energy requirement, which obviously rules out 'breaking the light barrier', manifests itself as a progressively increasing *inertia* as the body is forced to move faster. In the case of a rocket, instead of the fuel energy being converted into faster motion, more and more gets converted into the *mass* of the payload. Quite simply, the rocket just goes on getting 'heavier' and so becomes harder and harder to shift.

Of course, one could attempt to exceed the speed of light in more subtle ways, say, by dispatching two rockets each at speed $0.9c$ relative to the Earth, but in opposite directions, thus apparently giving them a relative velocity to each other of $1.8c$. However, a brief allusion to the transformation equations 2.2 and 2.5 gives a different picture. In arithmetic 0.9 plus 0.9 might make 1.8, but in relativity theory they make 0.995! That is to say, when the additional time dilation and length contraction are allowed for, each rocket actually measures their relative velocity to be only $0.995c$ rather than $1.8c$. Moreover, however many smaller rockets are fired ahead of these two at $0.9c$, the sum of relative velocities can never exceed the speed of light.

For these reasons it is often said that nothing can travel faster than light. Strictly speaking this is not true. Only material objects cannot be accelerated through the 'light barrier'. There is no known reason why there cannot exist superluminal bodies, provided they are *always* superluminal, i.e. cannot be slowed to less than light speed. Indeed, such bodies (in the form of microscopic particles) have been actively searched for by experimental physicists in the last decade or so. They have even been given a name – *tachyons*. So far no tachyons have been found. If they ever are found, it is expected that they would only interact with ordinary matter in an uncontrollable way, otherwise it would be possible to use them to send messages. This would then be the cause of an apparently unresolvable paradox, because it may be shown from the theory of relativity that tachyons can travel backwards in time, so that their use as a signalling device would facilitate communication with the past. One could then construct a booby-trapped device which would destroy itself by a coded signal sent into its past, thereby removing the possibility of sending the signal in the first place – an obvious contradiction!

Leaving aside tachyons, the nature of the barrier set by the speed of light is best expressed by saying that no *physical influence* can travel faster than light. One consequence of this is the impossibility of constructing a truly *rigid* body. This can

Fig. 2.7. Length contraction effect. From the garage the five-metre car appears shrunk to three metres when travelling at 80 % light speed. It will easily fit into the four-metre space. From the car it is the *garage* which is shrunk (to about two-and-a-half metres) and the car will not fit. The outcome is shown at the bottom. The rear of the car continues moving into the garage until it knows the front has stopped. The message cannot get through faster than light, so the car gets squashed up well inside the two-and-a-half metre space. Both views of events are consistent. However hard the chassis is made, no material in the universe can withstand this squashing.

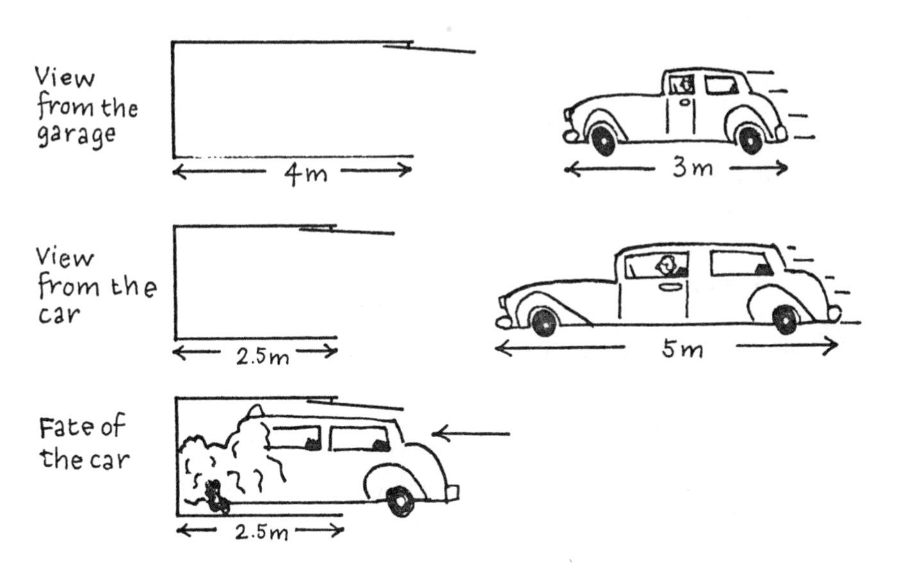

be illustrated by an amusing 'paradox' concerning a man who has a five-metre car but only a four-metre garage. Being a physicist he realises that by driving fast enough into the garage the length of the car may be reduced below four metres by the length contraction effect mentioned on page 44. Having taken a couple of turns round the block to work up a speed of about $0.8c$, the length of the car viewed from the reference frame of the garage is less than four metres. As soon as the car enters the garage, the automatic door closes and traps it!

From the viewpoint of the driver things appear to be somewhat different. Travelling inside the car, he naturally notices nothing unusual about its length. However, he does notice that the *garage* appears to be unusually flattened – in fact it is shrunk by exactly the same contraction effect, to a length of about two-and-a-half metres. Too late to stop, he realises his blunder: a five-metre car will *not* fit into a two-and-a-half-metre garage. Which view is right – a foreshortened car easily fitting into the garage, or a long car not fitting into a short garage?

As usual in relativity, *both* pictures are correct. The alternative stories are reconciled by taking into account what happens to the long car when it reaches the end of the short garage. Obviously it crashes into the end wall (which should be very robust). But this alarmingly violent event of a near-luminal velocity impact does not *stop* the car in the usual sense. The front of the car, which is abruptly arrested by the wall, obviously stops, but the rear of the car cannot know of this until the shock wave travels down the length of the car. As no influence, including the shock wave, can travel faster than light, the rear end of the car will have to wait at least 1.7×10^{-8} seconds (car time) to know of the existence of the wall. During that time it will have travelled, at a speed of $0.8c$, a distance of four metres. The result of this is that the car is compressed up into a length of a mere one metre, which easily fits into a two-and-a-half-metre garage!

The moral of this story is that, however rigid your car is made, however stiff and incompressible the material, there is always a certain amount of squashing possible at near-light speeds. In the next chapter it will be explained how one of the results of this relativistic squashing is to shrink up whole stars to nothing.

2.5 The new four-dimensional space–time

The initial response of many people to the seemingly preposterous ideas of time dilation and length contraction is one of instinctive antipathy. Our everyday concepts of space and time, strongly rooted in intuition and experience, are profoundly shaken by the relativity revolution. For some laymen, refuge is taken in outright scepticism.

In reply to this scepticism it should first be mentioned that by any standards the special theory of relativity is not a recent novelty. The original paper by Einstein was published in 1905, and within a very few years the theory had been accepted by the scientific community. Since then it has become one of the foundation stones of the edifice of modern physics, with implications which pass far beyond the rather trite considerations of high speed railway trains and rockets. In fact, the time dilation effect has been verified directly, both at the subatomic level and macroscopically by clocks being flown around the world. But more importantly, the novel principles involved in relativity have to be incorporated into all branches of physics. For example, the theory of gravitation and the laws governing the structure of the atom and the subatomic particles must be made consistent with the ideas of Einstein. When this is done many new effects, such as the splitting of certain atomic spectral lines, which at first sight seem to have little connection with relativity, are found to be in agreement with experiment.

One of the most spectacular confirmations of the theory arises from the rewriting of the laws of mechanics to satisfy the relativity principle. As mentioned in section 2.4, the light barrier results in the conversion of energy into mass, to prevent a body from reaching forbidden superluminal velocities. This conversion can also operate the other way, and the effect is contained in perhaps the most famous of all Einstein's equations

$$E = mc^2 \qquad (2.7)$$

This equation shows that an enormous amount of energy E may be obtained from a small amount of mass m, because the right-hand side of equation 2.7 contains the square of the speed of light, c, which is a very large number. Thus a mere one gram of matter can liberate about 30 million kilowatt hours – enough energy to supply the power requirements of the average home for many years. The conversion of mass to energy explains the (formerly mysterious) energy source of the sun, and also occurs in a more startling fashion during the explosion of the atomic bomb.

In addition to the impressive experimental confirmations of special relativity, the theory has great aesthetic appeal because of the additional symmetry and unification that it brings into theoretical physics. Many mathematical expressions are at once more elegant when recast in accordance with the new principles. In large part this is due to the unification of space and time which is suggested by taking a closer look at the theory.

At the outset of this discussion of the special theory of relativity it was remarked that the space and time model of Newton (or neo-Newtonian space and time) would need to be modified to accommodate effects such as time dilation and length contraction. We may obtain some insight into the new kind of space and time structure which results by the following considerations.

In Newtonian space and time it is assumed that spatial lengths and temporal intervals are independent of the motion of the observer or system. That is, rod lengths and clock rates do not depend on the relative motion of these objects and on who it is that observes them. In contrast, the theory of special relativity requires that the length of an object contracts along its direction of motion, and its temporal progress expands. Now we have seen from the train experiment that a moving object which is extended in space is extended in time also – events at the ends of the train regarded as simultaneous to a passenger are spaced a time interval apart to the observer on the embankment. This suggests that it is more accurate to regard an object as in general having extension in *both* space and time. One could regard the time dilation and length contraction effects (heuristically) as a reduction in spatial extension appearing as an increase in temporal extension. It would then be more appropriate to imagine an object as having unchanging *space–time* extension, with the *projection* of this extension on to both space and time as a varying proportion dependent on the relative velocity of the object. This projection might be expected to be similar to projections in ordinary space, where a rod of fixed length may appear to be foreshortened by orienting it so that its aspect along the line of sight appears shorter. The true length of the rod is related to the projected lengths in perpendicular directions by Pythagoras' theorem

$$l^2 = x^2+y^2+z^2 \tag{2.8}$$

x, y and z being the projected lengths in three perpendicular directions and l being the true length of the rod.

Now inspection of the transformation equations 2.2 and 2.5 reveal (with a little algebra) that although x and t are not invariant for all observers, the combination $x^2-c^2t^2$ *is* invariant, i.e.

$$x^2-c^2t^2 = x'^2-c^2t'^2$$

If we consider motion in three spatial directions rather than just in the x direction, this invariant interval, call it s, is given by including y^2+z^2 as well:

$$s^2 = x^2+y^2+z^2-c^2t^2 \tag{2.9}$$

A comparison of (2.8) and (2.9) shows that indeed the space and time projections may be combined together in a type of Pythagorean way, provided the time interval is multiplied by the speed of light, c, which, being a velocity, allows the conversion of an interval of time into an interval of space. Because c is so large, a very small time interval is 'worth' a lot of space. For example, one second of time is equivalent to 300 000 kilometres of space!

The fact that it is only by combining space and time intervals together that a fixed quantity s, the same for all observers, may be formed, suggests that space and time should really be considered as joined to form a unified *space–time* of four dimensions. The properties of this four-dimensional structure were first discussed by Hermann Minkowski (Russian, 1864–1909), so the space–time of special relativity is sometimes referred to as Minkowski space. It is a space in the mathematical sense of the word. It must not be supposed from this that space is really four-dimensional, or that time is really a form of space. The theory of relativity simply recognises the fact that the properties of space and those of time are closely interwoven, and separate models of each cannot be constructed.

The combination of space and time may often be made intuitively easier by drawing a space–time diagram or map. Most people are familiar with ordinary two-dimensional space maps, where longitude is plotted against latitude as shown in fig. 2.8. A path such as that drawn on the map might be the route of a road or river. A point on the map is the location of a place on the surface of the Earth. In analogy a map may be drawn to represent space–time. Naturally it is not possible to draw a four-dimensional map on a sheet of paper, so the diagram will be limited to time plus one dimension of space only (e.g. the x direction). In place of latitude, the horizontal lines should now be envisaged as a rigid rod at rest in the reference frame considered, each line being the position of the rod at successive times. The vertical lines ('longitude') measure the distance from the end of the rod. The path drawn on the space–time map is the path followed by a moving particle as time passes on. In the case shown the particle is first at rest in the frame considered, then moves a little to the left, after which it accelerates to the

Fig. 2.8. Space–time map. A point on the map is an event – occurring at a particular place and time. A path on the map is a history – a sequence of events, such as the contortions of a moving particle. 'Longitude' marks a place at all times, 'latitude' marks a single moment at all places.

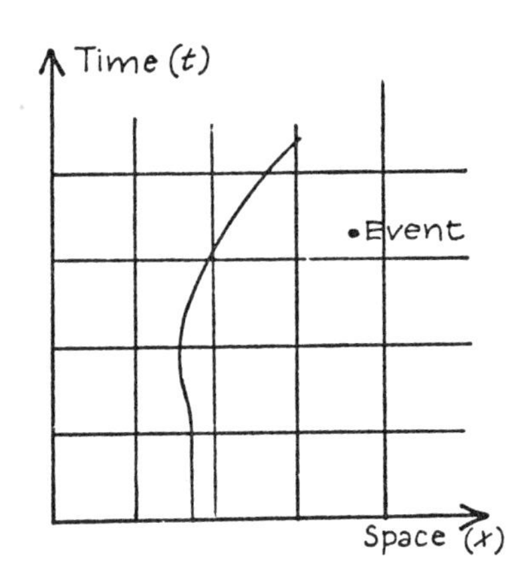

right. On a space–time map therefore, a path may be envisaged as the *history* of a moving point particle, usually called its *world line*. A point on this map is the location of an *event*, i.e. a certain place at a certain time.

Space–time maps may be used to emphasise the differences between Newton's and Einstein's models of space and time. There is no reason why Newtonian space and time should not be combined together into a space–time of four dimensions except for the fact that it provides no new physical structure but is merely a technical exercise. This Newtonian space–time is illustrated in fig. 2.9*a*. It possesses a natural slicing into space sections at constant time. All points on a given space section are simultaneous and make use of the same universal Newtonian time. This natural slicing is a consequence of the fact that we have merely artificially stitched together Newton's space and time, and then simply decomposed them again. The frame may be regarded as fixed in absolute space (or the ether if you like). All observers, irrespective of their state of motion will agree on

Fig. 2.9. Newton versus Einstein's space–time. The horizontal slices shown in (*a*) represent all of space at one moment; events *A* and *B* occur at the same time. In Newton's model these slices are the *same* for all observers, irrespective of their motion. In contrast, (*b*) shows Minkowski's model of Einstein's space–time. The map is only drawn correctly for one particular observer with a certain uniform motion. However, all observers agree on the light paths (oblique lines). All particle histories through *P* must lie inside the light lines. Events *A* and *B*, which lie in the 'outside' region, have no definite time order and cannot causally influence each other, or *P*, whereas *Q* is unambiguously *later* than *P*.

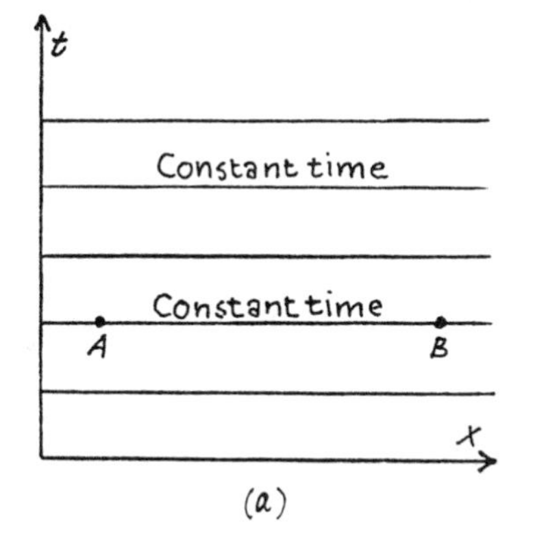

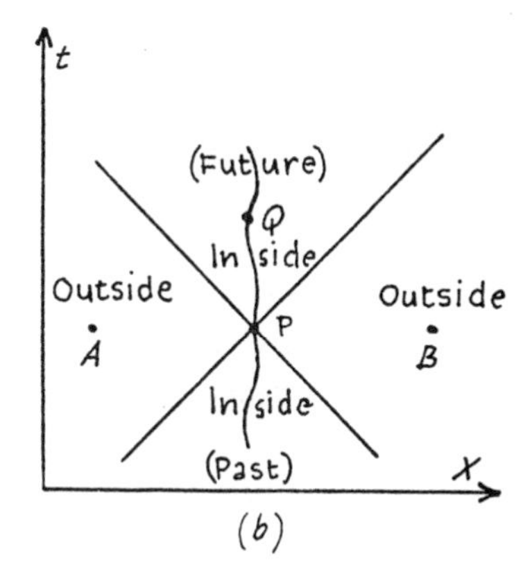

the form of this map, because they all make use of the same space and time.

In contrast, the Minkowski space of special relativity may not be thus decomposed. There is no natural slicing into space sections at a constant time because different observers disagree about simultaneity and length scales. However, a different type of decomposition of space–time emerges naturally from special relativity and is depicted in fig. 2.9*b*. The sloping lines are the paths of light rays travelling through a particular event p to left and right at constant velocity c. By choosing the units of length and time to be the same (for example centimetres and light centimetres, respectively) these lines slope conveniently at 45°. Now on the map which is illustrated, the lines marked x and t will only be drawn correctly in one reference frame. A different observer will wish to draw them with a different slope because of his different view of space and time. However, the light lines at 45° *will* be the same for *all* reference frames because all observers measure the same speed of light. These light lines therefore naturally divide Minkowski space into two regions – the 'inside' and the 'outside' (see fig. 2.9*b*). The path (world line) of a material particle must always be on the inside because it moves slower than light. Two observers in different states of motion will see different paths, but always inside the light lines. It is obvious from fig. 2.9 that the immovable nature of the light lines fixes every event inside them to a definite past and future relation to the apex event p. In contrast, the order relative to p of events outside the light lines is not fixed. An example of the reversal of the time order of events was given in the discussion of the fast-moving trains on page 33. However, the reversal of time order does not imply reversal of cause and effect. Because all known influences (particles, light signals, etc.) can only travel at, or less than, the speed of light, events in the outside region which have no definite time order relative to p cannot affect p in any way. Of course, if tachyons exist, they can travel in the outside region, and thereby reverse cause and effect. It is for this reason that paradoxical situations with tachyons can arise.

The difference between events which lie inside and outside the light lines is reflected in the value of s^2 given by equation 2.9. On the inside, the time part of the interval from p, c^2t^2, is greater than the space part x^2 (or $x^2+y^2+z^2$ in the full four dimensions). It follows that s^2 is negative. On the other hand, in the outside region it is the spatial contribution which is greater and s^2 is positive. Along the light lines themselves s^2 is zero, which means that four-dimensionally speaking the distance along a light ray is zero however far in space the light travels. These peculiar properties of the interval s^2 arise from the presence of the minus sign in front of c^2t^2 in equation 2.9. It means that distances cannot just be measured on a space–time map as they can on an ordinary space map, because the scale changes with the slope of the lines. The geometry of Minkowski space is therefore somewhat unusual. For example, it is possible for a vector in Minkowski space to be both parallel and perpendicular to itself!

In spite of its strange features virtually no physicists now seriously doubt the validity of the theory of relativity within the context of the situations described. If Newtonian theory were always correct, many phenomena which are very well understood in great detail would be quite without explanation. Nevertheless, the limitations of special relativity have been recognised for many years. In 1915 Einstein published a second great intellectual adventure, predicting that the special theory was in fact just that – an approximation of a general theory of relativity that took account of the effects of *gravitation*. That is not to say that special relativity is *wrong*, but merely that it is an approximation valid in the limit that gravitation is negligible; just as Newtonian mechanics is not wrong, but an approximation valid in the limit of small velocities.

Far from returning us to the cosy concepts of the pre-relativity era, the general theory of relativity leads to a world where things are stranger still.

3 The asymmetry of past and future

3.1 The meaning of time asymmetry

What is the cause of *change* in the universe? Why do some physical systems remain more or less the same while others evolve, transmute or decay, and what is the fundamental *nature* of change? An atom (at least its nucleus) stays relatively immutable as it follows its career of interactions and relations with its surroundings; day follows night with seemingly unending regularity. Yet motor-cars rust and fall apart, the wind and rain slowly erode the mountainside and human beings grow old and die. Why do terrestrial clocks need winding up to prevent them from stopping, but astronomical clocks – the pattern of the days, months, years – run on without assistance?

The quest for understanding *why* and *how* some things around us change while others do not is a long, controversial and disorganised chapter of scientific history. Because of the sheer diversity and complexity of the vast numbers of physical systems in our environment, the question of their temporal evolution has been mainly dealt with separately as it arose in each branch of science. Consequently, in thermodynamics and biology, information theory and statistics, mechanics, electrodynamics, cosmology and many more, this question has been tackled using the mathematical tools and language of the trade. Where these subjects have overlapped, disagreement and controversy on the question of time and change has arisen between their practitioners. Much of this controversy is actually avoidable if common concepts are clarified at the outset and physical questions are separated from philosophical ones. Perhaps the greatest misconception of all surrounding the problem of why things change with time is the confusion between time as it enters into the laws of physics and time as it enters into the human mind.

In chapter 1 a sharp qualitative distinction was drawn between the human perception of space and that of time. The most elementary experience of time is that of *uni-directional activity*, sometimes thought of as a flux of flow *of* time, sometimes as a movement of conscious awareness *in* or *through* time. There is no comparable experience for space. Yet the mathematical developments of Newtonian mechanics, Maxwellian electrodynamics, special and general relativity and quantum theory have only emphasised in varying degrees the similarity in structure between space and time. Nowhere in the development of theoretical physics has any need been discovered for a flowing, moving time. The world of relativity is described by a static four-dimensional map. Thus, the time which enters into the equations of theoretical physics seems to lack that essential quality of human psychological time. In chapter 7 this strange fact will be examined in detail, and some of the arguments which have been advanced in favour of the startling proposal to relegate the flow of time to the status of a psychological illusion reviewed.

Even disregarding the psychological one-way movement of time, there still exists a distinction between past and future. A good way of illustrating this distinction is to use the analogue of a motion picture film. Suppose a film is taken of some 'everyday' sequence of events; for example, a match burning itself to charcoal and smoke. The film consists of a set of frames – still photographs – and may be regarded as a physical model of the real world. Suppose that the film strip is cut up into separate frames, which are then shuffled and stacked in a pile. An individual is then given the task of sorting out the collection of frames into the correct sequence. Then even if the individual had not seen the actual match-burning event taking place, with sufficient care there would be little difficulty in *correctly ordering* the frames. This is because the condition of the match *changed* so there exists in this case a *correct* sequence in the sense that one, and only one sequence of frames displays the change which occurred in the real world.

Now suppose that the experiment is repeated with a motion picture film taken of the swing of a pendulum. Once again, many possible sequences of frames are incorrect, but this time there is not a single unique sequence that is a possible, correct description of the real world. For example, any sequence producing a motion picture film which displays a pendulum performing familiar oscillations will, if arranged in *reverse* order (or if the film is run backwards) display a pendulum performing equally familiar oscillations. Of course, in general, if the individual in the experiment had witnessed the original events, he could make a decision between the two possible sequences of frames (for example, he may have noted that the pendulum was started from the vertical towards the right, the reversed sequence on the film showing initial action to the left). But what is important here is not whether the reverse sequence of events *did* happen, but that it *could* happen, being perfectly consistent with the laws of physics and 'everyday' circumstances.

One way of expressing the different experiences of the individual in the foregoing two cases is to say that the former sequence of events is *asymmetric in time* whereas the latter is *symmetric*. It is worth noting at the outset that 'symmetric' doesn't necessarily imply periodic when used in this sense. For instance, an object which falls towards the sun from a distant region of space, then swings around in a close orbit and passes out again for good, does not move periodically but does move symmetrically in time because its motion is *reversible*. This means that if it passed along the trajectory in the reverse direction, it would still be displaying a quite unremarkable behaviour, perfectly consistent with the laws of physics.

In contrast to periodic and other symmetric phenomena which would cause no consternation if their reverse phenomena were encountered, asymmetric processes have the property that their reversals, if encountered, would appear to be miraculous. A piece of twisted charcoal and a cloud of hot smoke spontaneously combining together to reconstitute an unburned match would certainly be regarded as a miracle.

We now come to a crucial point. The asymmetry in time illustrated by the former collection of motion picture frames is *not* a property of time itself, but a structural property possessed by the collection of frames and, because the film is a model of the real world, a property also possessed by the real physical system (the match, smoke, etc. in this case). An explanation of time asymmetry in the universe should therefore be sought not in the structure of time itself, but in the structure of the universe, which seems only to produce asymmetric sequences of events in the same time order.

A failure to distinguish the property of time *asymmetry* – a property which happens to be possessed by the world we live in – from the previously discussed *flow* or movement of psychological time – a property which (psychologically) seems to be possessed by time itself – has contributed to generations of confusion and misconception about the *origin* of time asymmetry. The structural property of time asymmetry is possessed by the collection of motion picture frames even when they are lying in a pile on the projectionist's table. It is not necessary to stick all the frames together and actually *run* the movie film in order to discuss this asymmetry.

Part of the reason for the confusion between these two distinct concepts is semantic. In physics an asymmetry is often represented by an arrow pointing one way or another. For example, the Earth's rotation defines a very useful asymmetry because it distinguishes the north pole from the south pole. Standing at

Fig. 3.1. The arrow of time. Many processes only proceed in one time direction. This direction (e.g. gathering rust on a car) is named as the future, and may be denoted by an arrow. The arrow indicates that the world is asymmetric, it does *not* denote motion through time, a psychological phenomenon of mysterious origin.

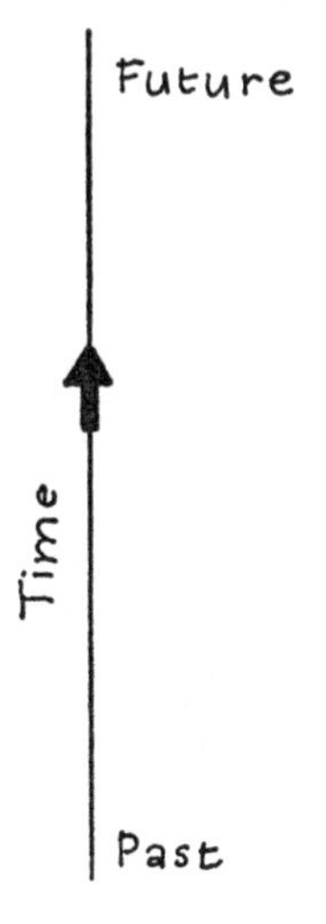

the north pole the Earth rotates anticlockwise beneath one's feet, at the south pole it is clockwise (recall the anecdote about the direction of bath water vortices escaping down the plug hole). For reasons of pure historical convention, presumably relating to the fact that mathematical navigation was invented in the northern hemisphere, it is customary to draw an arrow on maps and charts pointing toward the north pole. Many compasses are also equipped with such an arrow. But the presence of the arrow pointing north on a ship's compass in no way implies that the ship is actually *sailing* north. The arrow might just as well be chosen by convention to point south. In the same way, the time asymmetry of the world may be denoted by imagining an arrow pointing one way or the other in time. Which way is purely a matter of convention. In practice, the direction of the arrow is chosen to point with its head towards the time direction in which matches are burned and its tail in the direction in which they are unburned. In place of naming the former direction north and the latter south, we call the former the *future* direction and the latter the *past* direction of time. Naturally this convention means that the arrow points also in the direction of the flow of psychological time. But just as the ship is not required to move northwards, so the presence of an asymmetry in time denoted by a future-pointing arrow does not require time to *flow* forwards from past to future. We may indeed have the impression that it does so, but this has no (superficial) connection with the time asymmetry. Many authors refer to the *arrow* of time or the *direction* of time and do not make the distinction between time asymmetry and the flow of time.

The time-asymmetric character of the everyday world is so much a part of existence that the task of categorising asymmetric phenomena at first sight seems bewildering. A most conspicuous source of asymmetric change is biological activity. As individuals we start our lives as babies, slowly changing as we grow up and then grow old, finally being subjected to the very sudden change of death. We never encounter beings who grow younger with time. To a large extent changes in our environment are also biological in origin. The evolution of culture, modification of cities and technological changes all stem from human social organisation. Then again, areas of land left untouched soon swarm with life as nature takes over. The slow evolution of the species themselves provides a good example of biological asymmetric change.

Much intellectual change has to do with the accumulation of *information.* Records of past events, but not of future events, accumulate everywhere. Libraries fill with books, coal seams are laid down, beaches fill with footprints. Many of the features of the terrestrial environment are records of this character. Of course, although total information accumulates, specific information becomes degraded. The tide washes the footprints away, for instance. The degradation of information is always one-way in time: a noisy telephone line never contributes to the conversation, it only reduces the quantity of information passing between the subscribers.

In the inanimate environment there is a diverse range of asymmetric phenomena which bring about change. Changes from ice to water or water to steam, for example; a piece of ice placed in a jug of boiling water will melt to form warm water. The reverse never seems to occur, with warm water spontaneously freezing in one place and boiling in another. A great number of one-way changes are *dissipative* in character. Disturbances of all kinds tend to spread out and decay. Heat flows out of hot bodies into the cooler environment, gases diffuse away in the air, wave disturbances on ponds spread outwards and disappear, currents of air such as the wind 'blow themselves out', the heat and light from the sun and stars flow away into the surrounding space.

In fact, time-asymmetric change appears to be a feature of almost every natural phenomenon. Actually, even processes which at first sight seem to be symmetric in time are often, when viewed over a longer time scale, mildly asymmetric. For example, a pendulum will eventually be damped to rest by friction and air resistance, unless it is driven by a motor – itself a dissipative device. Even the motion of the Earth round the sun is subject to the incredibly minute viscous drag of the tenuous interplanetary medium. A rough way of summarising matters is to say that time asymmetry is a feature of all macroscopic activity.

It is a remarkable fact that the nature of change in the greater part of these examples can be understood in basic outline from the analysis of just one particular branch of physical science – thermodynamics. Although this discipline was originally devised to discuss the flow of heat between systems and the performance of heat engines, modern thermodynamics, now understood at a microscopic level in terms of statistical mechanics, covers such diverse topics that it includes aspects of almost the entire range of macroscopic (everyday size) physical phenomena. Moreover, recent, exciting advances in highly non-equilibrium statistical mechanics are providing a thermodynamic basis for the understanding of life. For the purposes of time asymmetry, biological change can be considered as a branch of thermodynamics. In addition, modern information theory can be formulated in terms of concepts which closely parallel those of thermodynamics and statistical mechanics, and the asymmetric degradation of information may now be considered an example of a general thermodynamic principle.

There is still some time asymmetry which does not appear to have a direct thermodynamic character. For example, radio waves leave a transmitter and spread out into space at the speed of light. The reverse process, in which radio waves arrive from all directions spontaneously from outer space, and converge on to a radio transmitter, never seems to be observed. Put more provocatively, a radio message is never received *before* it is sent, only after. An understanding of this asymmetry, and that of other wave motion, cannot immediately be obtained from a study of thermodynamics.

Cosmology provides a forum for the discussion of large-scale cosmic change. We live in an expanding universe, which evolves in its gross features with time. At the other extreme, the submicroscopic world of elementary particles possesses a strange occupant called the K^0 meson. This particle only lives for about 5×10^{-8} second after which it decays into three other particles. The curious fact is that the reverse process, in which the three particles reconstitute a K^0 meson, does not (unlike all other particle processes) follow precisely the time reverse of the decay.

The K^0 meson decay therefore defines a preferred orientation in time.

The subject of cosmology will be dealt with in chapters 5 and 6. K^0 mesons will not be discussed further because although their behaviour is intriguing, it seems unlikely that they exercise any very profound effect on time asymmetry in general. In this chapter the nature and origin of the asymmetry in the phenomena which admit a thermodynamic description will first be discussed, after which some attention will be given to the propagation of waves, in particular, electromagnetic waves.

3.2 Irreversibility and the second law of thermodynamics

As mentioned in the previous section, the laws of thermodynamics were originally constructed in order to describe the performance of heat engines. The so-called first law of thermodynamics provides a theoretical foundation for the fact that heat is a form of *energy*. Like all energy, it may be converted from one form to another. The steam engine is an admirable device for converting heat energy into mechanical energy, whilst the electric heater converts electric energy into heat. In all cases the total energy is conserved, and the first law simply expresses this fact.

Heat resides in a body in the form of molecular motion. Increasing the temperature causes the motion of the molecules to become more energetic. This motion is really very rapid. A typical molecule of air at room temperature is moving at several thousand metres per second. It follows that when two bodies at different temperatures are brought into contact, the rapidly moving molecules of the hotter body communicate some of their energy by collision to the molecules of the cooler body. Heat therefore conducts from a higher temperature to a lower one. After a while, both bodies will reach a more or less uniform temperature, when they are described as being in thermal *equilibrium*. Further observation would not reveal any subsequent transfer of heat (if the system is isolated from additional sources of heat).

This illustrates a general principle that, left to itself, a system never transports heat from a lower temperature to a higher. The spontaneous flow of heat is always from hot to cold. This is sometimes expressed by saying that the heat transfer is *irreversible*. Naturally, the change can be reversed, and heat returned from a cold body to a hot one, if some outside mechanism is used. For example, a refrigerator expels heat from its interior to the outside world, but only by the irreversible expenditure of some sort of external energy supply required to make the refrigerator operate.

The common experience that heat flows out of hot bodies and dissipates away in the cooler surroundings is an expression of one form of the *second* law of thermodynamics. It is manifestly a time-asymmetric law for it forbids the reverse process of heat passing from cold to hot. Since its formulation in terms of heat flow, it has been realised that the second law is of much greater generality and describes 'irreversible' time-asymmetric phenomena of many varieties.

In order to widen the second law to encompass other varieties of irreversible behaviour, a new quantity called *entropy* has been invented by physicists. The precise definition of entropy is a mathematical one, but it may be given various physical interpretations. One useful view of entropy is as a measure of *disorder*. A system which possesses a great deal of structure and order has

Fig. 3.2. Second law of thermodynamics. The first law regulates the *quantity* of energy in the form of heat, the second law regulates its *organisation*. Lord Kelvin (British, 1824–1907) expressed the second law by forbidding heat to flow spontaneously from hot to cold bodies. Thus the hot water can melt the ice, but the ice cannot boil the water. The time order of events is always from left to right as shown. The second law is of much greater generality. It is the most comprehensive regulator of natural activity known.

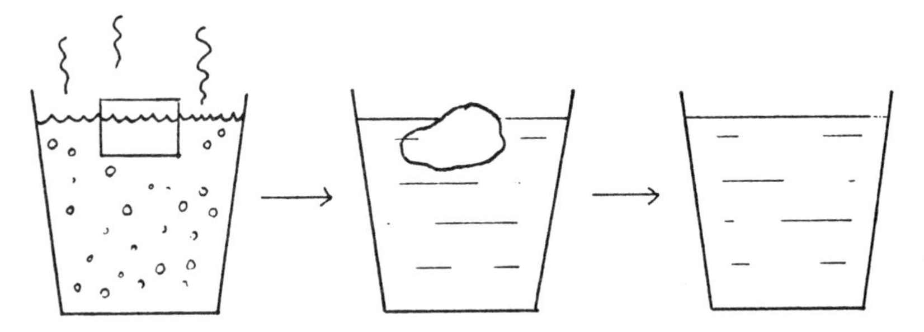

a low entropy. Conversely, high entropy systems are disordered and chaotic. For example, the entropy of a system which consists of a cold body placed next to a hot body is somewhat lower than the entropy of the same system with the two bodies in equilibrium at the same temperature. The reason for this is that the heat content of the system is more ordered when it is arranged to reside mainly in the hot body than when it is spread uniformly throughout the system. Put another way, there is a greater degree of structure in the former situation.

This suggests that the second law of thermodynamics be restated in the form that *the entropy of a system will never decrease.* It is necessary to restrict this statement at the outset to *isolated* systems, that is, systems sealed up in heat-proof containers. Clearly, if outside systems are permitted to come into interaction with the system of interest, the entropy may be decreased, say, by using a heat pump to transfer heat from a cold to a hot body. However, the second law does say that in the wider system, including (in this case) the heat pump, its energy supply, etc., the total, overall entropy always increases (or at best remains unchanged). A summary of the law is that the entropy of the universe can never go down.

Using the entropy concept, the condition of *equilibrium* can be identified with *maximum entropy.* Any change which occurs to an isolated system tends to increase its entropy. When equilibrium is finally established, no further change occurs, and the entropy cannot increase any more; it is a maximum.

It is also possible to relate entropy to *information.* If a physical system is in a highly ordered condition, it will require a large amount of information to describe it; or, looking at it another way, it will contain a lot of information. In contrast, a disordered system contains little information. One obvious example is the arrangement of letters on this page. By ordering them carefully into the correct sequence, information is contained in the form of words, sentences, etc. A jumbled and disordered collection of letters would provide little information for the reader. Information can thus be identified with *negative entropy,* or negentropy as it is sometimes called. When entropy increases, information is lost.

One advantage of formulating the second law of thermodynamics as an entropy law is its greater generality. A good illustration which will be frequently discussed in this book does not involve a heat transfer at all. Consider two different gases, call them *A* and *B*, sealed up inside a container which isolates them completely from the outside world. The container, which is depicted in fig. 3.3, consists of a box divided in half by a screen. In the left-hand section is placed a mixture of 90% gas *A* and 10% gas *B*. On the right is 90% *B* and 10% *A*. At a certain moment, the screen separating the two halves of the container is removed. After a short while the molecules of gas, which are moving randomly at great speed, diffuse among one another and become intermingled. Eventually, after a fairly short period of time, the two gases *A* and *B* will be more or less uniformly mixed up, with 50% *A* and 50% *B* on both sides. This process is clearly asymmetric in time, for we would not expect to encounter a

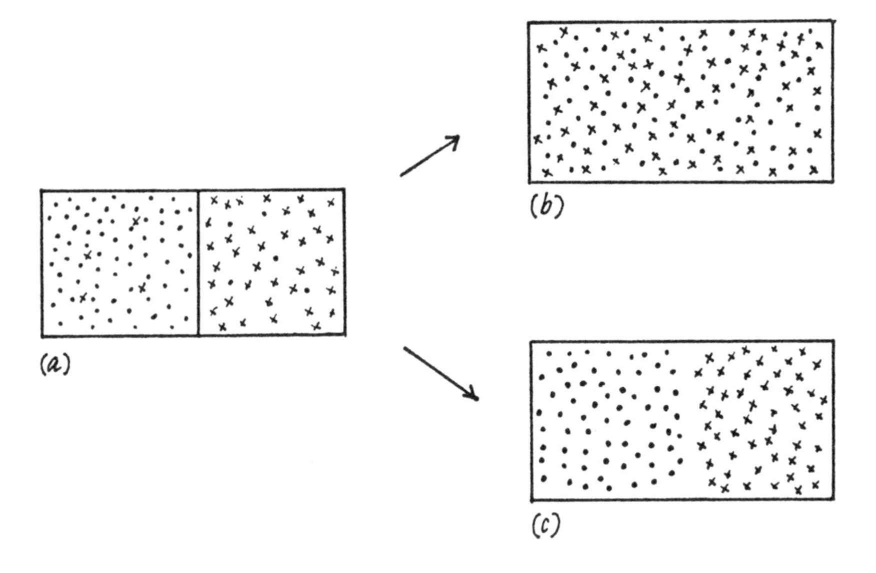

Fig. 3.3. The law of entropy increase. Gas *A* is denoted by dots, gas *B* by crosses. When the membrane is removed from the box the gases mix. Arrangement (*a*) is more ordered than (*b*), and so has a lower entropy. It also requires more information to describe it (we must state the mix on both sides of the membrane rather than throughout). A spontaneous decrease in entropy (a self-organising isolated system) such as (*c*) never seems to occur.

flask filled with a mixture of two gases in which the gases spontaneously unmixed themselves and retreated to opposite ends of the flask. The mixing process is well described by the law of entropy increase because the entropy of the original configuration, with the gases neatly separated out, is clearly more ordered (and contains more information) than the chaotic mixed state which results when the membrane is removed.

To generalise from this example, we have the following principle of nature: order tends to give way to disorder.

In human terms this principle is familiar. It is much harder to achieve a high degree of order and structure than to reverse it. A house is easily demolished to a pile of bricks, but requires great care to build up again. There *are* systems in which structure seems to appear naturally, and which at first sight apparently contradict the entropy law. Biological systems tend to evolve to more complex structures; crystals forming out of a liquid possess more ordered arrays of atoms than the liquid, and so on. However, a careful examination of these processes reveals that the total entropy of the system *and its surroundings* always increases. For example, biological activity only continues by virtue of the increase in the entropy of sunlight, which provides the energy source of all terrestrial life. A plant or animal sealed up in a box soon dies, conforming to the principle that, in isolation, order collapses into disorder.

In 1866 the first attempt to *explain* how order gives way to disorder was made by Ludwig Boltzmann (Austrian, 1844–1906). When the concept of entropy was first introduced into thermodynamics, the atomistic theory of matter was not very well developed. Entropy was in fact defined in terms of *macroscopic* quantities, like the temperature and pressure of a gas in equilibrium. Then the work of Rudolph Clausius (German, 1822–1888) and James Clerk Maxwell around the mid nineteenth century attempted to describe the differences in the states of gases by the differences in the *arrangements* of their constituent molecules. The molecules themselves were assumed to move as little particles obeying Newton's laws of mechanics. By mathematically analysing the effect of the collective motions of vast numbers of identical molecules, properties like the temperature and pressure of a gas in equilibrium were accounted for. With this microscopic view, the pressure of a gas is explained as the combined force due to the tiny impacts of all the individual molecules as they rush around and collide with the walls of the container. The temperature is understood in terms of the *speed* of molecular motion. The hotter a gas becomes, the faster its constituent particles fly about. The total heat content of the gas is the combined energy of all this motion (plus perhaps some internal rotation and vibration energy of the molecules themselves).

Boltzmann extended this theory of molecular motion to non-equilibrium situations in an effort to describe mathematically how a system could progress on its own from an arbitrary initial state to the equilibrium condition. This one-way in time process lies at the foundation of the time asymmetry of the physical world. Although Boltzmann studied the properties of a very special model representing a gas confined in an isolated box, investigations of more elaborate models have not revealed any fundamentally new principles regarding the time asymmetry.

For any given *macroscopic* state (with a certain distribution of temperature, pressure, etc.) of a gas in a box, there are in general many different arrangements of the positions and motions of the individual molecules consistent with it. Some states can be achieved by more arrangements than others. For example, there are more ways of arranging for the gas to be spread evenly throughout the box than when the entire amount is confined to one small region of the box. Also, there are relatively few ways of arranging for all the molecules to be moving in the same direction, whereas the numbers of arrangements in which the motions are all haphazard is vast. Thus, the more orderly the state, the less the number of alternative arrangements available. High entropy states can be achieved in many more ways than low entropy states. In this molecular theory of gases, there is one state which can be achieved by vastly more microscopic arrangements than all the others. This is the maximum entropy state, with the greatest disorder. Equilibrium is therefore identified with the state which can 'most probably' occur if the molecules are distributed at random.

The essence of Boltzmann's work on the progress of a gas towards equilibrium, contained in his so-called H theorem, was to combine the Newtonian laws of mechanics (which describe the molecular motions) with an assumption about how randomly the model gas seeks to rearrange itself.

Molecular rearrangements take place when the individual molecules *collide* with each other. The effect of the collisions can be envisaged as a *reshuffling* of the microscopic configuration. If this reshuffling is sufficiently random, it is fairly clear how the gas passes from a state of comparative order and low entropy to the highly disordered equilibrium condition. It is precisely because there are so many more disordered microscopic arrangements than ordered ones. It is simply an atomic counterpart of the familiar experience that a pack of cards arranged in a careful sequence will, when randomly shuffled, almost always pass to a highly disordered sequence. The chances of shuffling an arbitrarily arranged pack of cards into suit order, for instance, is obviously incredibly small.

Boltzmann therefore made a statistical assumption about the nature of the molecular collisions. This amounted to supposing that the motions of molecules about to collide is the same whether the collisions take place or not. That is, because the molecules don't 'know' a collision is going to occur, their motions are not affected (correlated) beforehand in any way. Of course, the motions *after* the collisions do depend on the fact that the collision has occurred.

Boltzmann called his assumption the *Stossahlansatz* or assumption of molecular chaos. Chaotically moving particles soon break up their orderly arrangement. Boltzmann's achievement was to supply a rigorous mathematical proof of this expectation for his model of a gas. In this proof he identified a quantity which he called H, and which depends upon the way in which the molecules are arranged. The H theorem proved that H could only *increase in time.* A close inspection reveals that H should be identified with entropy. The H theorem was thus hailed as a direct atomistic expression of the law of entropy increase. It apparently encapsulated the mechanism by which thermodynamic systems behave asymmetrically in time. This was undoubtedly one of the really great achievements of theoretical physics. The only trouble was that Boltzmann's H theorem was marred by a fundamental and devastating paradox, which has recurred in one form or another repeatedly in a century of controversy which followed.

3.3 The paradox of reversibility

There is no doubt at all that a molecular theory based on Newtonian mechanics alone cannot provide a proof that the entropy of an isolated system will always increase. The reason for this is disarmingly simple: Newtonian mechanics is symmetric in time. What this means is that any motion of atoms which proceeds in accordance with the Newtonian laws of motion has a *reverse* pattern of motion which is equally in accordance with these laws. Every collision, every trajectory of Boltzmann's model atoms is reversible. The mechanics upon which the whole theory is founded fails to make a distinction between one direction of time and the other. As asymmetry cannot be derived from symmetry, if Boltzmann's H theorem states that entropy increases asymmetrically in time, on the basis of Newtonian mechanics alone, it must be wrong. It makes no difference to this objection that Newtonian mechanics is now known to be an incorrect description of atomic motion, which must be described by quantum mechanics. Nor will relativistic considerations help. Both of these modern modifications equally fail to distinguish a privileged direction of time. Clearly, if Boltzmann had proved that entropy could only increase, he must have slipped in the time asymmetry somewhere in addition to the mechanics.

The first objection to a purely mechanical 'explanation' of the law of entropy increase was published by J. Loschmidt in 1876. Boltzmann's theorem apparently stated the following: choose any state of a gas, let some collisions occur, get a new state. The entropy of the new state cannot be less than the original state. Loschmidt disproved this result by finding states for which the entropy did decrease. These states are simply the *reversals* of the foregoing final states. For suppose by some magic all the molecular motions of a higher entropy state were simultaneously reversed; then the gas would be sent 'backwards' to its original, lower entropy, state. The reason for this is that if every individual molecular collision is reversible, then the total gas motion must be reversible. The conclusion to be drawn from Loschmidt's objection is that not every state of a gas will lead to a subsequent entropy increase.

Another objection was raised by E. Zermelo and also hinges on the time symmetry of the underlying laws of mechanics. Henri Poincaré (French, 1854–1912) had proved a general theorem about isolated mechanical systems obeying reversible laws of mechanics. This theorem stated that such a system would return an infinite number of times to within arbitrary closeness of any given state. It follows from this theorem that if an isolated box of gas is initially in a low entropy state, it must eventually return to a nearby (low entropy) state. There is no way in which the system can return again to a low entropy state without contradicting Boltzmann's H theorem.

Poincaré's theorem is so remarkable that it is amusing to dwell on some of its consequences. A loose way of expressing the theorem is by saying that, in a totally isolated system, anything that can happen, will happen, and infinitely many times! Consider the situation in my living-room if it were completely sealed off from the rest of the universe today. After a sufficiently long time my table would spontaneously rise to the ceiling. The flowers on the table, long since dead, would bloom again. I myself would be reincarnated time and again. Occasionally all the atoms of air in the room would rush into one corner. Any given situation would recur again and again, infinitely often. The only trouble is, I would have to wait a very long time indeed for these strange events to happen. The so-called Poincaré recurrence time – for example, the time between reincarnations – is perhaps the longest conceivable relevant time scale. It may be underestimated as 10^N where N is the number of particles which make up the system. For a human being and immediate environment N is perhaps about 10^{26} atoms. The Poincaré number $10^{10^{26}}$ should be carefully digested. It is one followed by 10^{26} noughts. Compare this to the one followed by a mere 33 noughts which scales the limit of quantum gravity in centimetres.

It scarcely matters whether $10^{10^{26}}$ is expressed in seconds or ages of the universe – for what are a dozen or so extra noughts compared to 10^{26} of them? Poincaré's theorem tells us that miracles can happen, but they are more rare than we could ever conceive.

Naturally, it can be objected that the total isolation of my living-room is a physical impossibility anyway, which is certainly true. At one time, however, the entire universe was regarded as a suitably isolated system for the application of Poincaré's theorem, and Boltzmann even conjectured that the present state of the universe was indeed a Poincaré miracle. We shall take up this and other outlandish novelties in chapters 5 and 6.

To return to the main theme of Boltzmann's H theorem and the reversibility objections thereto, the conclusion of these objections is now clear. In addition to the laws of Newtonian mechanics, Boltzmann also made the assumption of molecular chaos. This *Stossahlansatz* cannot always be correct. The way in which it fails can be understood by considering more carefully the microscopic motions of the molecules. Focus attention on a small group of such molecules rushing around haphazardly in the box, bouncing off each other continually, always moving from one pattern of motion to another. Most of the time the collective behaviour of this little group will be unremarkable. For example, more or less as many molecules in this group will be moving to the left as to the right. Very occasionally, however, a chance alignment of motions will occur, or some cooperative behaviour will result, quite by accident during the course of events. Such slight statistical overbalances will bring about fluctuations in the gas. These fluctuations can actually be observed in a real gas in the form of something called Brownian motion. A microscopic particle placed among the shifting pattern of molecules does not remain at rest, but gets knocked about because of the slightly unequal molecular bombardment on its different surfaces. The motion of the particle is observed to be a random zigzag path.

A simple statistical theory shows that the probability of these chance alignments of molecular motions becomes rapidly smaller as the number of molecules involved increases. The chances of a whole box of gas showing a distinct fluctuation (e.g. all the molecules rushing simultaneously into one-half of the box) is astronomically small. It is not, however, strictly zero. It follows that entropy *can* decrease, and a disordered gas *can* spontaneously organise itself into a more structured, ordered configuration, but in practice such an occurrence is *overwhelmingly improbable.* To take a specific example, consider once more the partitioned box of gas discussed on page 66. According to Poincaré's theorem, eventually the time will come when all the molecules of type *B* on the left will just happen to be moving simultaneously to the right, and those of type *A* on the right moving leftwards. If the partition is removed at this instant then the gases will neatly separate out, as depicted in fig. 3.3*c*. But the probability that the removal of the partition will coincide precisely with the moment of this extraordinarily rare event is so small that for all practical purposes it may be considered zero. Everyday physical systems contain so many molecules that their increase of entropy is virtually assured and may be considered to be a law of nature.

On the basis of this new statistical interpretation of molecular chaos, the H theorem may be reconciled with the reversibility objections. *If* the system is in a relatively ordered, low entropy state, almost certainly it will soon rearrange itself in a less ordered, higher entropy state; it does not *have* to. On the other

Fig. 3.4. An isolated box of gas cannot distinguish past from future. The entropy may change from a maximum as chance re-arrangements sort the molecules into a collectively remarkable state. These are the dips in the graph. Big dips are overwhelmingly less frequent than small ones. The graph shows clearly that the entropy changes are *not* oriented preferentially in time.

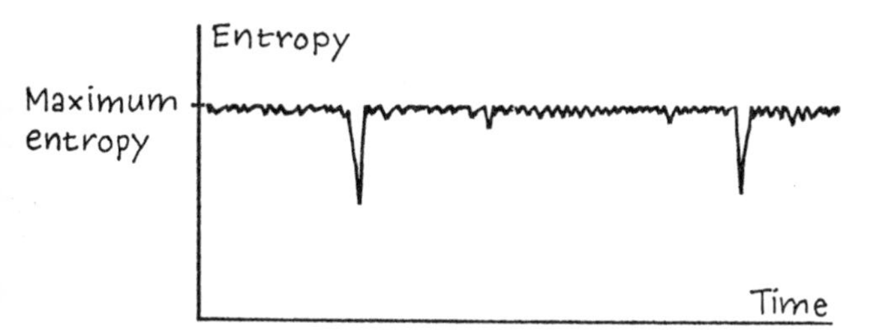

hand, by symmetry, the system must also very probably have reached that low entropy state from a higher entropy state in the past. That is to say, a randomly selected, low entropy state is very, very probably near an entropy *minimum.* This can be easily appreciated by considering the long time behaviour of an isolated box of gas, depicted in fig. 3.4. Most of the time the gas is close to equilibrium at maximum entropy, with the molecules all disordered and spread out evenly, moving in all directions. Occasionally a minor fluctuation interrupts the calm equilibrium and the system spontaneously acquires a certain amount of structure, which is soon smoothed out again by random collisions. On exceedingly rare occasions, a really large fluctuation occurs and the entropy drops dramatically, as, for example, when all the molecules collect in one-half of the box. A low entropy state picked out at random would almost certainly lie near the bottom of one of these dips, because there are so many more little dips than big ones. That is to say, if the gas gets itself into a remarkable state, it is far more likely that this state represents the *limit* of its extraordinary behaviour, rather than a half-way stage to some still more preposterously unlikely self-organisation. But at the bottom of a dip, the entropy curve is clearly symmetric in past and future. The *Stossahlansatz* is a good assumption at this point, but it applies equally going forwards or backwards in time.

The utility of Boltzmann's H theorem is that it describes *how* a model gas can pass from a low entropy state to an equilibrium state. It does *not* explain though, why this always happens in the same time direction – past to future. The time asymmetry of Boltzmann's model has disappeared!

3.4 The hypothesis of the branch systems

An inspection of fig. 3.4 immediately shows that a permanently isolated box of gas does not behave asymmetrically. Its long time behaviour causes it to visit low entropy states on rare occasions, but its entropy decreases as often as it increases. The time direction in fig. 3.4 could be taken to be either way.

Obviously real systems are not like this. Time asymmetry is a fact of life, so in the real world boxes of gas must differ in a very fundamental way from Boltzmann's model. What is this difference?

Taking a more elaborate model than Boltzmann does not fundamentally alter matters. The answer to the question comes from asking how a *real* system gets into a low entropy state in the first place. According to the Boltzmann model, an arbitrary low entropy state is almost certainly an entropy minimum, having been created by a fluctuation from equilibrium. To expect this to apply to real systems is absurd. If I encounter a half-obliterated sand castle whilst walking on the beach, I conclude that there was once a *whole* sand castle. Yet according to the reasoning applied to a Boltzmann-type system, it is supposed to be far more likely that just prior to my encounter with this structure, it was actually *less* well formed rather than more. However, although I know that the disturbing effect of the wind and tide will certainly be responsible for eventually completely smoothing out the sand castle (the smooth state is the equilibrium state), I do not thereby conclude that the wind and tide must *also* have been responsible for miraculously *forming* the structure in the first place. The beach is not a totally isolated system; someone *built* the sand castle a short while previously. The system achieved its low entropy, ordered condition, not by an exceedingly rare fluctuation of the wind and tide, which just happened to sweep the sand into a castle-like shape, but by *outside interference*.

Return to the example discussed already, with the box containing gases *A* and *B*. The original arrangement of a 90%, 10% mix could be achieved by patiently waiting for a uniform mixture of gases to execute a 'miraculous' fluctuation from a 50%, 50% mix, after which the screen could be inserted. But in the real world, such an inefficient procedure would never be entertained. Instead, one simply opens up an empty box and inserts the gases in the desired mix. Real systems receive their orderliness by manipulation from the outside world, not by random fluctuations.

It is useful to think of a system as being formed anew as a result of this external interference. In many cases, systems actually *do not exist* prior to some outside action such as, for example, when an ice cube is placed in boiling water. It is obvious that in the real world *all* systems must have been formed by the wider environment at some stage.

In order to distinguish real systems, which form in low entropy states by separating off from the rest of the universe, from permanently isolated, Boltzmann-type systems, the German philosopher Hans Reichenbach (1891–1953) introduced the term *branch system* to describe the former. All real systems are branch systems of one sort or another. Often there is a whole hierarchy of branches, each one dependent on another for its ordered structure. In chapter 6, all of these branches will be traced back to a common 'trunk'.

A branch system which has just been formed, behaves asymmetrically in time because it has been tampered with from

Fig. 3.5. Real systems are branch systems. If I encounter a falling egg (*c*) I quite reasonably predict that it will soon *after* be a broken egg (*d*). I do not 'retrodict' that it was also broken *before* (as in *a*). I think it rolled off the shelf (*b*). If the system was isolated, as in Boltzmann's model system, then (*a*) is far, far more likely than (*b*). In the *real* world, (*b*) followed by (*c*) is usual while (*a*) followed by (*c*) is miraculous. (*a*) can only change into (*c*) if all the disorganised little atomic jiggles of the egg plus floor cooperate to form up a whole egg which shoots upwards in an organised motion, thereafter to drop back to its destruction. (*a*) $\rightarrow$ (*c*) $\rightarrow$ (*d*) is time symmetric, (*b*) $\rightarrow$ (*c*) $\rightarrow$ (*d*) is asymmetric. To find the origin of the asymmetry we must ask the question: 'how did the egg get on to the shelf?'

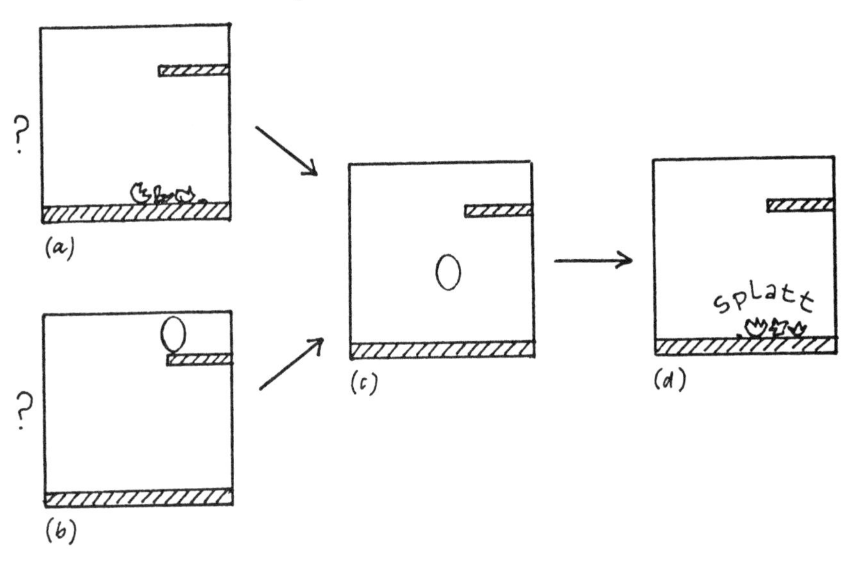

the outside. The asymmetry is in the tampering, not in the system. So far so good, but the appreciation of this vital step doesn't tell *which way* the asymmetry is directed, only that there will be one. Nor are we assured that every time a system is set up as a branch system, or interfered with from without, the asymmetry will always be in the *same* time direction.

These remarks are well illustrated by the mechanism depicted in fig. 3.3. The box, when divided by the screen, is really two separate systems. When the screen is removed it becomes a single system. It is the action of removing the screen which is an act of tampering by the outside world.

Consider the following experiment. Start with 90 % gas *A* and 10 % gas *B* on the left and 90 % *B*, 10 % *A* on the right. Remove the screen for a few moments and then put it back. What is the result? It is of course common expectation that the end product is about 50 % *A* and 50 % *B* in both sections of the container. Now do the experiment again starting with a 50 %, 50 % mix both sides. Remove the screen and replace it. What result? Common expectation now dictates that there is no change; still 50 %, 50 % at the end.

This is an intriguing paradox because the outside tampering in this case seems to be time-symmetric – the screen is taken out and then reinserted. Film the sequence and play it in reverse; the screen comes out, the gases *separate*, the screen goes back. This is *not* the correct description of the second experiment even though the initial condition of the gas in both cases is *macroscopically* identical. The film shows a transition from a 50 %, 50 % mix to a 90 %, 10 % mix, but the experiment shows a 50 %, 50 % mix with no change. Why does the film when run forwards correctly describe the first experiment, but when run backwards fail to correctly describe the reverse (second) experiment? Furthermore, why should one expect 1000 similar systems set up with a 90 %, 10 % mix to *all* change to a 50 %, 50 % mix and not, for example, for some to produce a 99 %, 1 % mix? In other words, why is the entropy change in all these systems always parallel?

As a first step to resolving the paradox, appreciate the difference in the microscopic condition of the gas at the beginning and end of the first experiment. At the beginning, when the screen has just been removed, it is overwhelmingly likely that the molecules are moving about chaotically. If this is so they will start to destroy the orderliness of the 90%, 10% mix. On the other hand, at the end of the experiment, when equilibrium is achieved with a 50%, 50% mix, the situation is very different. Looking back in time, the disordered molecules have just come from a more ordered state. In 'reversed time' they are by no means moving chaotically, but according to a definite plan which connects them with the ordered 90%, 10% mix. In contrast to this, in the second experiment described above, with the 50%, 50% mix remaining unchanged, there is no such difference between the two temporal ends of the experiment.

The time symmetry of the external manipulation is therefore an illusion. We have to enquire how the 90%, 10% mix was made. If this was from a chance fluctuation, then we are just as likely to end up with a 90%, 10% mix and to start off with a 50%, 50% mix, provided of course the screen is moved in and out at random. However, if the gas has been *put* in the box in its orderly state *before* the experiment, the time symmetry is broken. Provided the condition of the gas at the outset of the experiment is a *random* one, then the subsequent increase of entropy is assured with high probability.

Several questions now arise. Why does the outside universe permit the 90%, 10% disequilibrium mix to be set up in the first place? Why do branch systems form in low entropy states? Furthermore, what is the origin of the initial *randomness* of the microscopic motions in these branch systems, so essential to the subsequent time asymmetry of their behaviour? These fascinating issues of why the universe is in thermodynamic disequilibrium in the first place, and how its microscopic constituents acquired their random motions is a contentious issue which properly belongs to the subject of cosmology, to be dealt with in chapters 5 and 6.

Meanwhile, there is some unfinished business. It has been assumed that the branch systems which separate off from the wider universe are therefore isolated systems. This is a fiction. For the purposes of simplicity the Boltzmann model was assumed to be sealed off totally from the outside world by an impermeable container. Such containers do not exist in the real world. Whilst certain materials are good insulators against heat flow, the atoms in the container wall will always be in contact with the outside universe, and influences from the wider environment will eventually percolate through the walls, transferring disturbances to the molecules contained therein when they suffer collisions with the walls. If nothing else, then the gravitational forces of the surrounding matter can never be eliminated. Moreover, most branch systems are not even enclosed in insulated boxes. So the question has to be faced as to what extent this residual, continuous interaction of the system of interest with the rest of the universe is relevant to the considerations already made about branch systems and time asymmetry.

At a microscopic level, the effect of these unpredictable disturbances is to destroy the *reversibility* of the system. This is

Fig. 3.6. Order into chaos. This usual transition may be inverted by a sufficiently dexterous individual who simultaneously reverses the motions of all the balls. The balls then return to situation (*a*) from situation (*b*). How do they 'know' their way back? The system is *closed*, so all the information about state (*a*) is still contained in the positions and motions of all the individual balls in state (*b*). However, if the walls of the billiard table are not rigid, but shake about, the delicate reversibility property is destroyed.

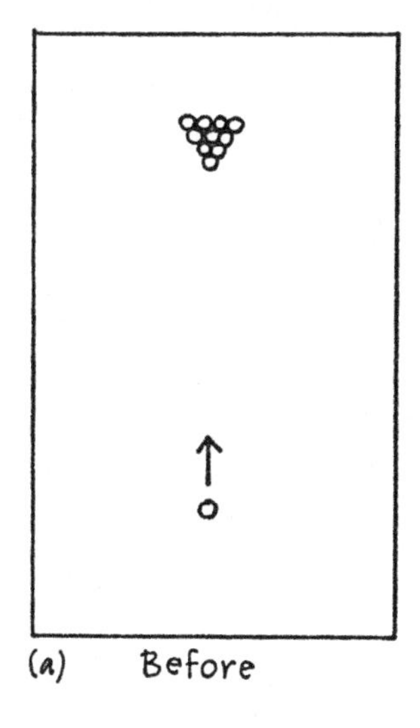

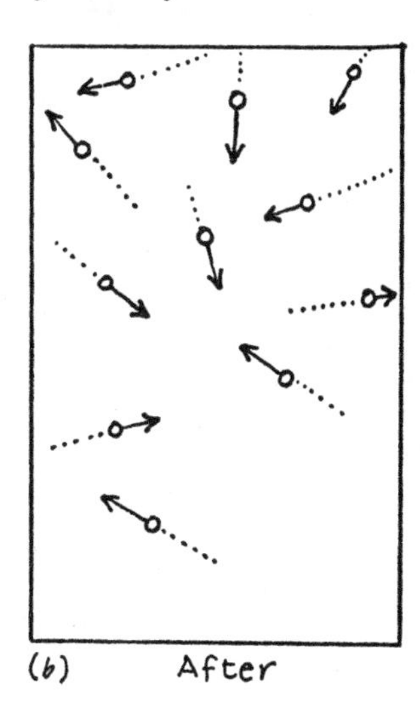

illustrated in fig. 3.6, which shows a billiard table and a set of snooker balls. The balls represent atoms of gas and the billiard table the container. Consequently, for the purpose of the illustration, friction will be neglected. At an initial moment, the balls are arranged as shown in (*a*), all confined to a neat triangle except for the cue ball, which is projected towards this little group. A few moments later, the situation is as depicted in (*b*) with a rather chaotic and haphazard pattern of motion as all the balls fly about in different directions, occasionally colliding, and spreading themselves about over the billiard table. This is another illustration of the general principle of entropy increase; the highly ordered initial state (*a*) has given way to the disordered state (*b*). As usual, to get (*a*) back from (*b*) is a very difficult process. It would be possible to do this if the motion of all the balls could be reversed simultaneously, say, by interjecting in the path of each ball an elastic screen in the precise way so that it would bounce back exactly along its previous trajectory. Provided the walls of the billiard table remain rigid then the balls would indeed end up in a neat little triangle again, with the cue ball racing away on its own down the table. That much is assured by the underlying time symmetry of the laws of mechanics.

At this stage, the effect of the outside disturbances may be incorporated by allowing the edges of the billiard table, which represent the walls of the gas container, to jiggle about slightly in a random fashion. Repeating the experiment, the cue ball is used to break the triangle. The subsequent configuration after a few moments is now different from (*b*), because each time a ball collides with the vibrating edge of the table it may get a little extra push, or perhaps lose some momentum. The new configuration will, however, appear similar to (*b*), with chaotic, haphazard motions spread out all over the table. The difference between this state and (*b*) is nevertheless revealed at once when the reversal procedure is carried out. Because of the perturbing effect of the jiggling edges, the balls will only travel back along their previous trajectories as far as the nearest edge. The collisions with the edges on the return journey will in general be only slightly different, but enough to change the whole pattern of motion. The triangular configuration shown in (*a*) is very highly ordered. Even the slightest disturbance to the reversed motions will result in balls arriving 'late' for vital collisions with each other and so on, and failing to position themselves precisely for the formation of the triangle.

The ability of a fully isolated mechanical system to return to its initial state through a reversed sequence of motions is sometimes picturesquely described by saying that the system 'remembers' its initial condition. The ability of the balls to find their way back again depends on their containing all the information necessary to build up their earlier ordered configuration. This information will be retained by the system as long as it remains isolated from the outside world. As soon as the random motion of the edges is introduced, this information passes out into the wider universe and disappears. To bring about a successful reversal in the latter case we should have to follow the effects of the collisions beyond the edges of the system, and arrange to reverse also all the minor perturbations which cause the edges to jiggle about. This slow 'loss of memory' in real systems is both irreversible and time asymmetric.

In information theory, random disturbances of this sort are referred to as *noise*. In every real system there will be noise due to the contact of the system with the outside universe. This 'cosmic noise' was not included in the discussion of Boltzmann's H theorem, or even branch systems, the time-asymmetric behaviour of which did not appear to depend on considerations of a tiny, continuous coupling to the wider environment, but only upon the sudden, violent interjections of the outside world in the formation process.

It is occasionally argued that the time asymmetry displayed by Boltzmann's H theorem applied to branch systems is an illusion. We only say that the 50 %, 50 % mix of the two gases in fig. 3.3 is less ordered than the 90 %, 10 % mix because we cannot view the motions of the individual molecules of the gases. The time asymmetry which results from the mixing process is a consequence of the *macroscopic* view of the gas to which human beings are restricted. Therefore, some people have maintained that the asymmetry is entirely subjective, a consequence of human limitations and not a fact of nature. It is alleged that the only 'true' asymmetry is that caused by the continuous random disturbances of the cosmic noise, because these cause a 'genuine' irreversibility at an atomic level.

Suppose a gas has just passed from a 90%, 10% mix to a 50%, 50% mix. An inspection of the end state would not enable us to infer that the gas was recently divided 90%, 10%, because *any* previous state (including the equilibrium one) would evolve after a short while into the 50%, 50% mix, as we have seen. At the level of macroscopic inspection, therefore, the information of the original state has been lost. However, at a microscopic level, the information is still there, contained in all the individual molecular motions, provided of course that the system is a totally isolated one. According to the alternative reasoning, this implies that a totally isolated system *never* reaches 'true' equilibrium, and there is no 'real' asymmetry. Only when the cosmic noise destroys even the microscopic information is 'true' equilibrium achieved.

As a response to this alternative point of view it has to be questioned whether or not the whole distinction between 'apparent' and 'true' equilibrium and time asymmetry is really relevant. Two unmixed gases will surely continue to mix whether or not there exists cosmic noise; what is required is an explanation of this phenomenon, and such is provided by Boltzmann's H theorem, together with an assumption about randomly produced branch systems. The existence of a 'true' asymmetry, such as might be attributed to cosmic noise, doesn't seem to have any very profound consequences for physics. Our own macroscopic knowledge of the world, which contains a strong feeling of asymmetry, even if it is only an illusion at the atomistic level, is well explained by the former mechanism. The objection to it is purely philosophical and seems to have little to do with physics.

3.5 Time asymmetry and wave motion

Until now this chapter has dealt only with the thermodynamic foundation of time asymmetry. Yet there are other important asymmetric processes which do not seem to depend *directly* on thermodynamics for their description, though an explanation of the asymmetry may connect with thermodynamic asymmetry at a more fundamental level.

One familiar example of non-thermodynamic time asymmetry is that encountered when a stone is thrown into a pond. The resulting pattern of disturbance is a set of circular waves spreading outwards from the point of impact towards the edges of the pond. The time-reversed situation, in which circular disturbances coherently generate themselves at the edges of the pond and collapse towards a common point of extinction, never seems to occur; at least, not spontaneously.

This wave-motion asymmetry turns up in many branches of physics. For example, in the propagation of radio waves. A radio message is always received *after* it is sent, never before. The radio waves pass *outwards* from the transmitter into the universe, never in the reverse direction.

The type of wave motion described here, where the disturbance travels outwards from the source, is called *retarded* wave motion by physicists because the disturbance reaches a distant point after a delay for propagation across the intervening space. The

Fig. 3.7. Retarded and advanced wave motion. A stone thrown into a calm pond causes ripples to travel outwards. These are called retarded waves (*a*). Equally possible, but never encountered, are advanced waves (*b*) where disturbances from widely separated regions of the pond cooperate to send incoming, organised wave motions towards the centre.

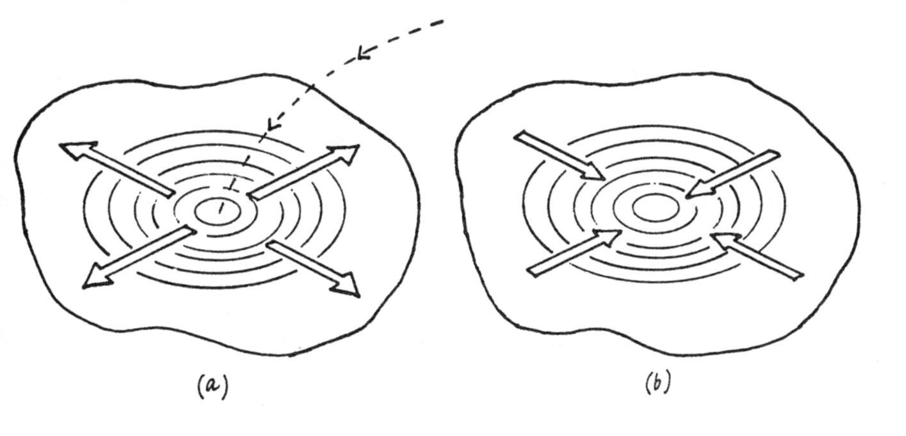

time-reversed situation, in which the disturbance passes a distant point *before* it collapses on to a source, is called *advanced* wave motion. The mystery is that the laws of wave propagation themselves make no selection between advanced and retarded waves. For example, if Maxwell's electromagnetic equations are solved, they possess both advanced and retarded wave solutions. The advanced solutions may be discarded by claiming that the conditions a long way off in space are not suitable for producing an incoming wave disturbance. But why not? Physicists have never been able to agree on a common answer to this question.

First let us take a situation where a clear answer does seem to exist. In the case of a pond, the system under discussion is finite in size. Proceeding in the same way as for the Boltzmann model of a gas, suppose we now have a model pond sealed off from the outside world, and also sealed off from thermodynamic processes such as viscous damping of the waves and so on which might complicate the issue. In this idealised system all types of wave pattern will occur, including, after a sufficient recurrence time, converging, or advanced, wave motion. Most of the time, however, the surface disturbance will be chaotic and disordered – a 'high entropy' condition – which we can think of as the equilibrium state of the pond.

This is in direct analogy with Boltzmann's system where eventually all types of molecular motion occur, but with the chaotic equilibrium state nearly always prevailing. Perhaps this close analogy between the two systems is not surprising. The quantum theory tells us that the motion of atoms has both wave and particle aspects.

A sealed pond behaves symmetrically in time, as does the sealed box of gas. However, a real pond is not sealed. If we throw in a stone from the outside we make a branch system, just as we do when we put an ice cube in a glass of boiling water. And once again, if the stone is thrown at random, then there will be retarded wave motion with overwhelming probability, because almost certainly any existing pond disturbance will not be poised on the brink of a remarkable fluctuation from its generally chaotic pattern of motion at just the moment when the stone is thrown.

Unfortunately, these considerations of branch systems break

down when the size of the system becomes *infinite*. If a box of gas is opened in an empty, infinite universe, the gas will evaporate away never to return. A radio wave, which travels out through unbounded space and does not encounter an 'edge of the pond', will likewise never return. This unbounded dissipation of waves and particles is a new type of irreversible time asymmetry which requires a new explanation. Clearly, such an explanation cannot depend upon *local* considerations. What must be explained is why the conditions in distant parts of the universe are such that converging radio waves and converging clouds of atoms never arise. The prospect of such an extraordinary event as radio signals travelling 'backwards' from the edges of the universe may simply appear ludicrous to the reader, and an explanation of their non-occurrence an empty academic exercise. Yet we shall see that the restrictions which rule out such bizarre events may place severe constraints on the type of universe we live in. Moreover, some model universes, which could bear resemblance to our own, may indeed permit these ludicrous happenings to occur from time to time.

An analysis of the origin of branch system formation, and the irreversible asymmetry of infinite wave motion have both led to considerations of the large-scale properties of the universe. Before convincing explanations of these processes can be given (in chapter 6), it is necessary to describe what is now known about the origin, structure and evolution of the universe. But first the nature of gravity must be understood.

4 Gravitation and the bending of space–time

4.1 Physics from a falling body

The relativity of uniform motion, which forms the foundation stone of both Newton's theory of mechanics and Einstein's special theory of relativity, depends crucially on the existence of *inertial frames of reference*. All systems moving with uniform velocity are considered as mechanically equivalent in these theories. Only by accelerative motion are physical differences felt *within* the moving system. According to Newton's second law on page 17 (and also Einstein's generalisation thereof) acceleration can only be eliminated and the special state of uniform motion achieved by removing the action of all forces upon the moving system. The existence of an inertial frame therefore depends on the ability to be able to acquire, at least in principle, a force-free state of motion.

Modern physics describes *four* forces of nature. The strongest – called in fact the *strong* interaction – operates at very short range, about 10^{-13} centimetre, between the particles in the atomic nucleus and is responsible for 'gluing' the nucleus together. Another weaker force – the *weak* interaction – is also a very short-ranged force between subatomic particles, responsible for, among other things, β-radioactivity. Neither of these two forces has any effect on the motion of *macroscopic* objects, and both were unknown when Einstein discovered special relativity.

The two remaining forces are electromagnetism and gravitation, both of which do act on large bodies. Inertial motion cannot be achieved while these forces are present. Consider how in practice one might determine whether a particular system was free of such forces. In an electric field, charged particles are accelerated, at a rate inversely proportional to their masses. By examining the behaviour of different types of particles, the existence of the electric field may be inferred. For example, positive and negative electrically charged particles will be accelerated in opposite directions in the field, and will move apart under its action. Electrically neutral particles (e.g. ordinary atoms, containing equal numbers of positive and negative charges) would not be accelerated at all. In addition, more massive particles with the same charge would be accelerated at a slower rate because of their greater inertia. A system freed from electric forces could therefore be achieved by making it electrically neutral, or very massive; in short, by reducing the ratio electric charge: inertial mass to a negligibly low value. The force-free nature of such a system could be directly verified by the above-mentioned observations of various charged and uncharged test particles.

In the case of gravity this strategy fails completely. An examination of the motion of test particles under the action of the Earth's gravity first reveals that all particles fall *downwards*. There is no known substance which falls away from the ground

Fig. 4.1. Motion under electric forces. Different particles released in an electric field move differently. The negatively charged sphere attracts the two equal positive charges with the same force, but the lighter one accelerates faster because of its lower inertia. The ratio of charge to inertia may be varied greatly, and even made negative (particle repelled) or zero (neutral particle unaffected). The presence of the electric field may always be inferred at a point in space, irrespective of the observer's state of motion, by releasing test charges of different varieties.

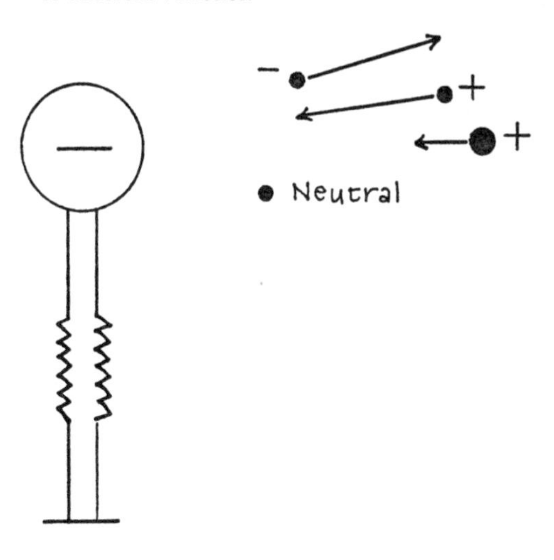

(like the imaginary selenite of H. G. Wells). Gravity always *attracts* between objects, never repels. Expressed differently, gravitational charge is always of the same sign, unlike electric charge which can be either positive or negative. Anti-gravity in the direct sense appears to belong only to science fiction. In spite of this, the presence of gravitational forces on a system would be apparent if different objects accelerated at different rates under gravity. The force-free state could be achieved if the ratio gravitational charge:inertial mass could be arbitrarily reduced.

The rate at which different objects fall may easily be determined by dropping them in the Earth's gravity. There is a story that Galileo actually performed this experiment from the leaning tower of Pisa. In any case he was the first person to discover the principle, as significant as the result of Michleson and Morley, that *all objects fall with equal acceleration.* Most people find this result intuitively implausible. It always seems that heavy objects should fall faster than light ones. However, heavy objects are

Fig. 4.2. Motion under gravitational forces. Different particles released under gravity move the *same* way. In (*a*) a heavy and light body dropped together hit the ground simultaneously. In (*b*) heavy and light bullets are fired with the same velocity. The path of the heavy bullet (large dots) is identical to that of the light bullet (light dots). These results are only approximately correct on Earth because air resistance interferes.

The presence of gravity *cannot* be inferred at a point in space by releasing different test particles. For instance, the curved trajectories in (*b*) could either be due to the downward pull of gravity, or the upward acceleration of the observer.

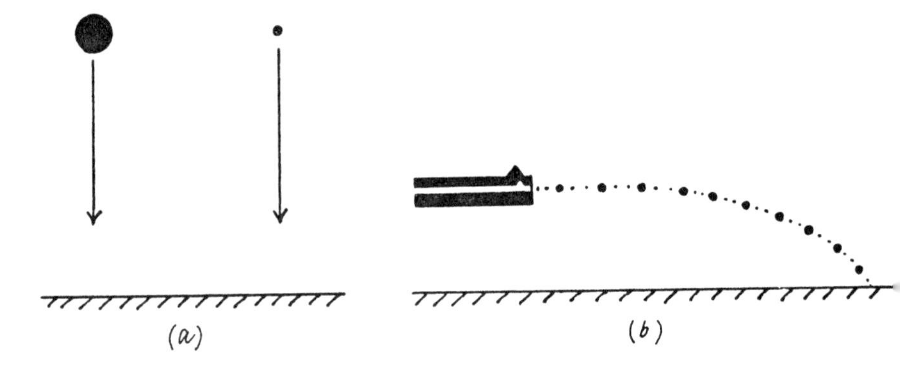

more massive, and so harder to shift. Galileo discovered that the two properties – being heavier and having greater inertia – always compensate exactly. Gravity pulls a rock more strongly than a pebble, but the pebble responds more readily. The result, easily verified by the reader, is that the rock and the pebble strike the ground together if dropped together (see fig. 4.2). Of course, a feather or a balloon will appear to contradict Galileo's principle, but this is due to the interfering effect of air resistance, and is unrelated to the nature of gravity. That was Galileo's achievement – to isolate the significant general characteristic of gravity from the irrelevant, though conspicuous, feature of air resistance.

Galileo's result has been checked more recently by Roland von Eötvös (Hungarian, 1848–1919) in 1889 and Robert Dicke in 1964 to an accuracy of one part per million million. Its content is best expressed by saying that the ratio of gravitational charge to inertial mass is *constant*, independent of the nature of the falling body. That is, gravitational charge and inertial mass are really *equivalent* physical properties of a body. For this reason Galileo's principle is generally known as the *principle of equivalence*. In its general form the principle of equivalence requires that all 'test particles' move along the same trajectory under the action of gravity.* Consequently there is no way in which an examination of the local behaviour of different test particles can reveal the presence of gravity. There are no such things as gravitationally neutral particles which are free from the effect of gravity, and against the motion of which our system of interest may be compared to see if it is force-free.

A very intuitive expression of the principle of equivalence is available if one recalls that the gravitational charge quantifies the force of gravity, while the inertial mass quantifies the forces of inertia due to accelerative motion. An identification of these two quantities amounts to a physical equivalence between gravitational and inertial forces. Such an equivalence is well known from experience. The centrifugal force on a rotating roundabout feels exactly like the force of gravity (only it acts horizontally). Indeed, it is frequently termed 'artificial gravity'

* 'Test particle' here means a particle which is small enough so that the effect of its own gravity on the motion can be neglected.

by space technologists who propose to utilise rotating space stations to reproduce a gravity of one g at their periphery, so that the occupants may work under normal conditions. The centrifuge is a method for producing very high gs by this method.

Conversely, the gravitational force of the Earth is indistinguishable locally from an accelerative effect. An individual sealed inside an opaque box (see fig. 4.3) would not be able to tell whether he was sitting at rest on the Earth's surface, or accelerating upwards at one g a long way out in space where the Earth's gravity is negligible.

Fig. 4.3. Principle of equivalence. The box accelerates upwards at one g a long way off in space where the Earth's gravity is negligible. The occupant feels his normal weight and cannot tell that he is not at rest on the ground. Other *local* observations are similarly normal. The plumb line hangs towards the floor, the bullet (which is really moving in a straight line at uniform speed) passes through the box on an apparently downward curved trajectory. In all respects the inertial forces due to the acceleration are locally equivalent to the force of gravity. Observations in a wider neighbourhood soon reveal the difference however. Gravity is due to the presence of nearby objects (e.g. the Earth) whereas inertial forces are not.

Note that the broken line could equally well represent the path of a light beam. This suggests that light is bent by gravity also (see fig. 4.5).

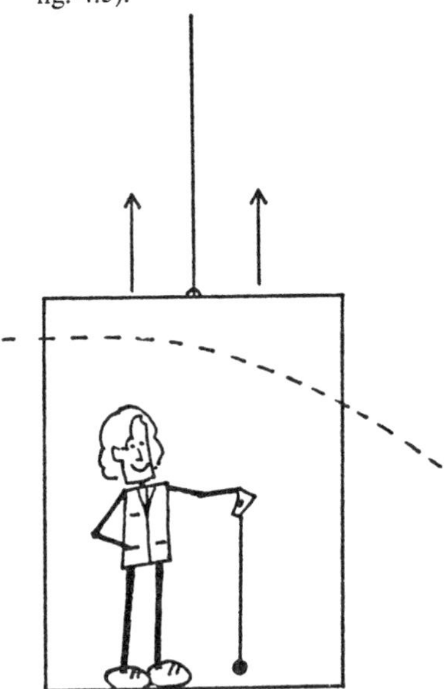

Just as an acceleration may be used to simulate a gravitational force, so too may gravity be *transformed away* by an acceleration. This is the situation experienced in a system which is allowed to *fall freely*. If the above-mentioned sealed box is dropped from the top of a cliff, the hapless occupant feels no sensation of gravity as he falls down with the box and contents. He, and other objects in the box around him, all fall at the same rate as the box, according to the principle of equivalence. Thus, his pipe, which has fallen from his mouth in the excitement of the moment, continues to hover a few centimetres in front of his face as they both fall at an equal rate. The view presented to the occupant inside the box is one of a gravity-free environment, with everything floating about, unaccelerated with respect to the box, and

Fig. 4.4. Freely-falling frames of reference. Local gravitational effects vanish (are 'transformed away') in free fall because of the principle of equivalence. In (*a*) all contents of the box fall downwards equally rapidly, so appear weightless to the observer. Moving objects, such as the bullet passing through the box, appear to travel along straight trajectories. (*b*) The orbiting astronaut is also in free fall, so appears weightless. His acceleration (due to the curvature of the orbit) locally transforms away the Earth's gravity. The latter is still present, of course, as the nearby falling meteorite knows.

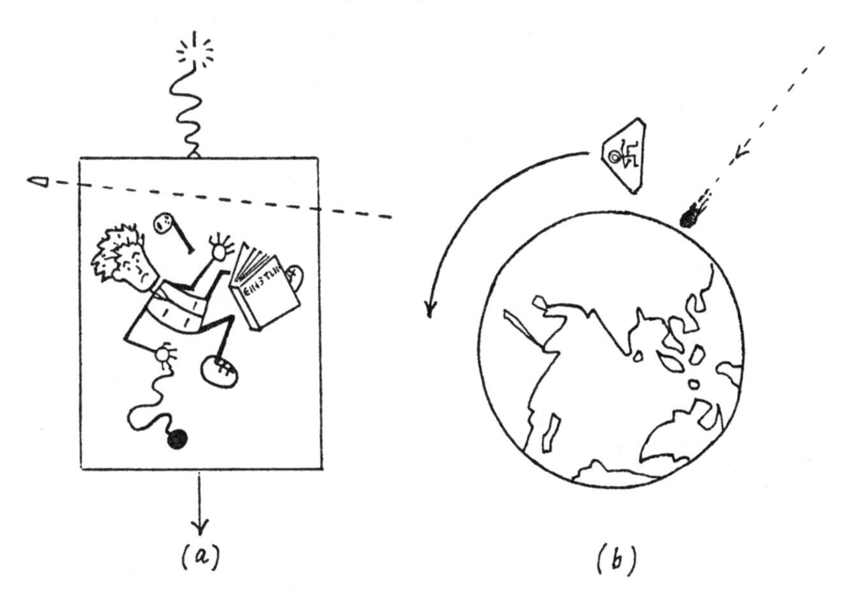

apparently weightless. The occupant cannot therefore deduce the presence of the Earth, or its gravity, while in a state of free fall. Of course, he will know all about the existence of the Earth when the box hits the bottom of the cliff and violent non-gravitational forces cause large relative accelerations in his body.

Such a situation described in the foregoing example must have been quite unfamiliar to (living) people when Einstein first drew attention to it. Today it is commonplace to see scenes of gravity-free environments inside space capsules. When the rocket motors of a capsule are not firing, it is in free fall, and the astronauts feel no gravity. The Earth's gravity is still present of course, and is appreciable even as far as the moon (how else would the moon remain in orbit about the Earth?), but the free-falling system cannot detect it. In a circular orbit 200 kilometres high, the gravity of the Earth is only about 6% less than at the surface, but it is transformed away inside the capsule by the accelerating motion of the capsule as it swings around along its freely-falling, curved path. Crudely speaking, the artificial gravity of its 'roundabout' motion round the Earth just balances the gravity of the Earth to give the weightlessness which is such a conspicuous feature of space travel.

The equivalence of gravitational and accelerative (inertial) effects should not mislead the reader into imagining that gravity is a type of illusion which depends only on the reference frame we happen to be using. The much-heralded equivalence principle is only true *locally*, which is why the restriction was made to a sealed box. Observations carried out over great distances can readily detect the existence of gravity. This is because gravity *varies* from place to place. Gravity may be thought of as a *field*, like the electromagnetic field of Maxwell, and this field is not uniform, but changes in strength and direction. Only in a sufficiently small region of space will the gravitational field be (approximately) the same throughout. For example, the Earth's gravity a long way off in space is far less than near the surface. A space capsule orbiting the Earth at 200 kilometres circles once every one-and-a-half hours or so, whereas the moon, slowly falling round the Earth in the feeble gravity 400 000 kilometres away, takes a full 28 days. There is thus a considerable relative acceleration between the capsule and the moon, even though within the small confines of the capsule there is apparent weightlessness. An astronaut who could see the motion of the moon relative to the capsule could therefore deduce the presence of the Earth's gravity (even if he couldn't see the Earth). Observations carried out over large regions of a gravitational field will reveal the existence of the field. The variation of the field from place to place is called the *tidal effect.* It is precisely the slight change in the moon's gravity across the Earth which causes the ocean tides. If the moon exerted a uniform force of gravity everywhere, it would have no effect on the oceans.

In the example of the falling box, tidal effects are negligible because the box is so much smaller than the Earth. Nevertheless, all the contents of the box are not really falling on parallel paths, but are actually falling towards the centre of the Earth. Consequently, these paths slowly converge, and if the box fell through a hole in the ground, they would all meet when the box reached the Earth's centre. Thus, although the falling occupant would be too startled to notice, the Earth's gravity is very slightly detectable as the debris around him is slowly drawn towards the centre of the box. Of course, the effect is small; falling from a cliff 100 metres high, two objects three metres apart would only move together by about a few hundredths of a centimetre.

To summarise, whereas an electric field may be detected with test charges at a point, only the *variation* of the gravitational field from place to place has physically detectable consequences.

To return to the subject of inertial frames, Newton was well aware of the equivalence of gravitational charge and inertial mass. Nevertheless, he assumed that if one considered a region *remote* from all sources of gravitational, electromagnetic and other forces, then the conditions would approach that of an inertial, force-free system. Inertial frames could then be constructed, at least in principle, at any point in the universe by observation and comparison with this distant system. The concept of an inertial frame still had a real significance, because it could be extended from the force-free region across the entire universe if need be. ('Constructing' an inertial frame does not literally mean building a rigid framework of metal rods or whatever. It is a mathematician's jargon for establishing a system of coordinates associated with a particular state of motion.) Thus, with a distant object moving uniformly in a force-free region of space, an observation of its motion from Earth would suffice to determine the rate of acceleration of a system at the Earth's surface, caused by all the forces, including gravity, which are acting on the system.

Newton's reasoning was unacceptable to Einstein because, quite apart from the fact that gravity is always present everywhere simply due to the background field of the universe, he knew from his theory of special relativity that comparison of a local state of motion (frame of reference) with a distant inertial system could not be carried out as Newton had supposed, even in principle. The reason for this is the equivalence of mass and energy, $E = mc^2$, which requires that light, a source of energy, also possesses mass, and so it will be affected by gravity in the same way as a material particle. The bending of light by a gravitational field was a central prediction of Einstein's theory, and was triumphantly verified observationally by Sir Arthur Eddington (British, 1882–1944) during an eclipse of the sun in 1919, when the bending of starlight by the sun was measured and found to agree with the theoretical value calculated by Einstein (see fig. 4.5). However, the fact that light is bent by gravity automatically

Fig. 4.5. Light is bent by gravity. The sun's gravity bends the light beam so that the distant star *A* appears from Earth to be shifted to position *B*. This shift can be observed and measured during an eclipse, when the moon conveniently blocks out the sun's glare, allowing the stars to be seen in daytime.

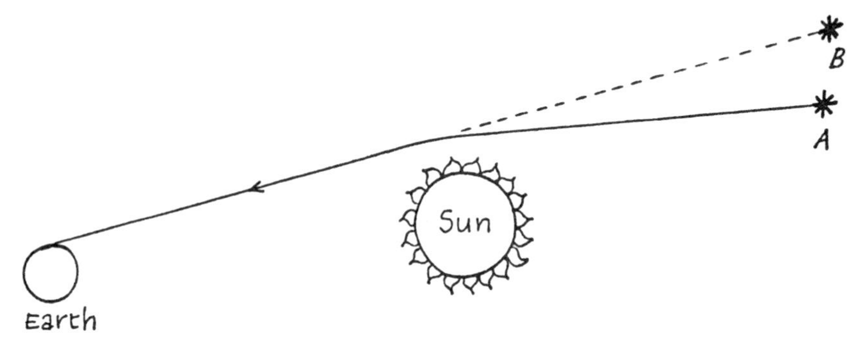

rules out the use of light signals to observe the motion of a distant inertial system from a region which is experiencing a gravitational field, because the observations are distorted.

As a result of deliberations such as these, Einstein came to the conclusion that the structure of space and time, then embodied in the principles of the new special theory of relativity, could not be separated from the considerations of gravity. He set about attempting to construct a new theory of gravity to replace that of Newton which had served so successfully for over two centuries.

4.2 The general theory of relativity and the gravitational distortion of space–time

When Einstein discovered the principles of special relativity, the two known forces of nature – electromagnetism and gravitation – had quite a different status in that theory. Electromagnetism was really the midwife of special relativity. The theory grew out of an attempt to reconcile the behaviour of electromagnetic waves, such as light, with the mechanical properties of moving bodies. As a result, Maxwell's theory of electromagnetism was automatically consistent with the principles of special relativity. Naturally, the *physical* interpretation in terms of Maxwell's ether had to be changed, but mathematically the electromagnetic theory was still correct. Such could not be said for Newton's theory of gravitation. This was based on an instantaneous action-at-a-distance force, which became a meaningless concept once the relativity of simultaneity, and the limiting nature of the velocity of light were realised. For how could a change in one body influence another distant body at the same moment when all physical influences are restricted to travel at most at light speed? And in whose reference frame was the 'same moment' to be reckoned?

In his search for a new theory of gravity, consistent with the principles of relativity, Einstein was guided by several clues. Firstly, in Maxwell's successful theory, the source of the electromagnetic field is electric charge, which does not change when viewed in different reference frames. In contrast, the mass of a body, which is the source of gravitation, does vary between reference frames – a particle gets heavier and heavier as it nears the speed of light. The type of field which Einstein was looking for was thus a more complicated field than Maxwell's. Whereas the latter can exert forces in different directions, the gravitational field should contain even more components. (Newton had suggested just one component of force, with the gravitational force always directed along the line joining the centres of gravity of two bodies.) The relationship between these many components was suggested on the basis of certain mathematical principles which are beyond the scope of this book.

Einstein also wished to incorporate the vital principle of equivalence at a fundamental level in the theory, and not to treat it just as a coincidence as Newton had done. By 1915 he had done this, and he published his new theory of gravity – a general theory of relativity – in final form in that year. To understand the general theory, return to the notion of a space–time map. In special relativity attention is focussed on uniform motion, which is represented by *straight lines* on the map. These straight lines form a privileged class of world lines which reflects the special status of inertial motion in that theory.

Fig. 4.6. (*a*) shows a space–time map, drawn by an observer standing on the ground, of a falling body (up and down are drawn horizontally by convention). In (*b*) is the corresponding map drawn by a falling observer. According to the principle of equivalence, gravity is locally transformed away by the acceleration of free fall, so that the path (world line) of the nearby falling body appears straight.

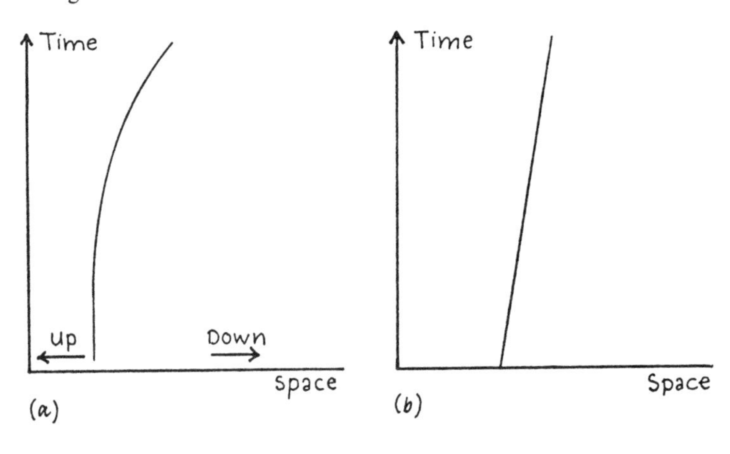

When gravity is present, inertial frames cannot be constructed. The world lines will bend around under the action of gravitational forces. However (ignoring for this discussion all other non-gravitational forces), *locally* it is possible to choose a frame in which the nearby world lines straighten out – the freely-falling frame. As we have seen, from a freely-falling system, the behaviour of nearby particles appears to be gravity-free, so that a space–time map of the region inside a small falling box is a good approximation to the special relativity case, with very nearly inertial motions realisable. Freely-falling frames thus replace the inertial frames of special relativity as the privileged class of motions. However, such frames may only be constructed locally. A space–time map representing a large region of space will show the small relative accelerations of the distant particles as viewed from one particular falling system, due to the tidal effects mentioned in the previous section. It follows that whatever falling system the region is viewed from, a map of the particle paths is bound to introduce a progressive *distortion* from simple straight lines as the region surveyed is made larger and larger. Since this distortion is the *same* for all particles, irrespective of their constitution (by the equivalence principle), this suggests at the outset that it might be more appropriate to regard the gravitation which is causing the distortion of the particle world lines to be a *property of space–time itself* rather than as some influence which is superimposed upon it.

It is possible to draw a space–time map in a more general way in such a fashion that the distortion of the particle paths disappears. That is to say, the map represents the view of free-falling observers *everywhere*, and not just centred on one particular local falling system. The nature of this generalisation is suggested by a comparison with ordinary terrestrial maps. There is a well-known systematic distortion on world maps which grows towards the outer edges of the map. For example, a map of the world shown in Mercator's projection depicts accurately only the equatorial regions of the Earth's surface. As one moves nearer and nearer to the polar regions, so the features of the map become progressively distorted. This distortion is particularly pronounced for Greenland and Antarctica, which become drawn out horizontally far in excess of their true proportions. The reason for this is well known, of course, it being simply due to the fact that the surface of the Earth is spherical, and it is not possible to represent a *curved* surface on a *flat* map without distortion. However, the distortion may easily be eliminated by drawing the map on the surface of a globe instead, which is an accurate representation of the geometry of the Earth. When this is done, the equator no longer has a special status on the map. What were drawn as *straight* lines on the flat map (e.g. lines of longitude) become *great circles* on the globe (great circles divide the surface of the globe equally).

These considerations suggest that the distortion which occurs on the space–time map is likewise due to the fact that in reality *space–time is not flat, but curved.*

Fig. 4.7. Why space is curved. A cloud of particles is dropped from rest above the Earth. They all fall towards the centre (drill lots of holes through the ground to allow this). The paths (world lines) on the corresponding space–time map all converge to this point at some later time t. Although on a small scale the *falling* observer sees nearly straight, parallel paths, the lines are curved gradually inwards to a point. This is reminiscent of lines of longitude on an ordinary map, which also converge to a point (at the poles). If they are drawn straight and parallel we never notice any distortion on the scale of a city, but on a world map (Mercator's projection) the distortion is considerable in the polar regions. This is because the Earth is round and the map is flat. These considerations suggest that space–time in the presence of gravity is really curved rather than 'flat'.

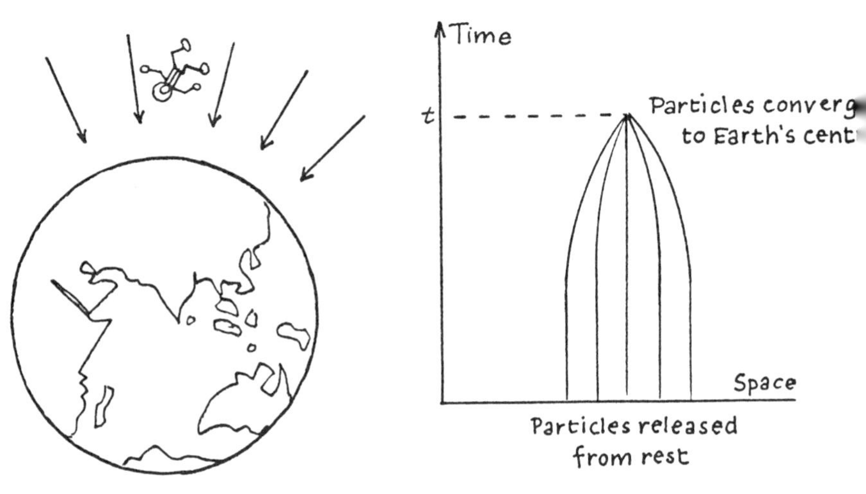

The idea of curved space–time may appear startling or even incomprehensible initially. Some of the properties of curved space–time are illustrated by considering the effect of this curvature on space and time separately. The possibility of curved space has been entertained by mathematicians for many years. The features of such spaces are best understood by comparison with flat space. When we talk of *flat* space, we mean a space (of any number of dimensions) which obeys the rules of geometry taught to every schoolboy. These rules taken together describe what is known as Euclidean geometry, after the Greek geometer Euclid, as mentioned before. When we do two-dimensional geometry at school, it is always carried out on a *flat* sheet of paper. If the paper is kept flat, then various familiar properties of geometrical figures may easily be verified. An elementary property of this nature is the fact that the three angles of a triangle add together to make two right angles. These rules, deduced from figures drawn on flat surfaces, will also hold in three-dimensional space if it is always possible to construct

Fig. 4.8. Three-dimensional space appears flat? Certainly to the accuracy of 'everyday' instruments the rules of geometry deduced from drawings on flat sheets of paper can also be applied without noticeable error to three-dimensional volumes (at least on Earth). For example, the length of an oblique path (*AE*) up a hillside (*ACEF*) can be calculated in terms of the gradient (angle *CAB*) and the obliqueness (angle *CAE*) by decomposing the figure into flat triangles and rectangles, as shown. Because plane geometry also seems to work for volumes of space, we say that space appears to be *flat*. More refined instruments show it is really slightly curved.

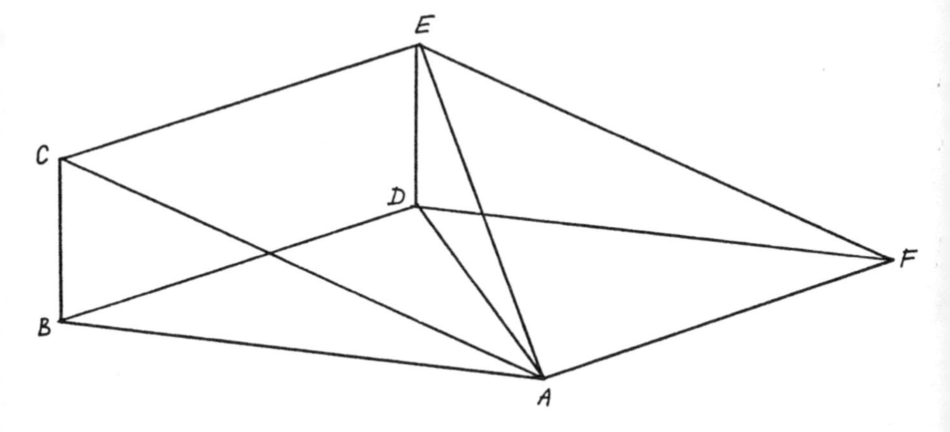

slices through space which are themselves flat. That such a slicing into flat planes is possible was never doubted by the Greek geometers, and is not doubted by many people even today. And indeed, when angles and distances in three-dimensional space are measured (it is not necessary to actually *draw* the figures) with theodolites and measuring tapes, it is found that to the limits of accuracy of these instruments, Euclidean flat plane geometry is obeyed (see fig. 4.8) near the surface of the Earth.

Now consider the possibility of a curved space. An example of such a space (mathematical space) is the surface of a curved sheet embedded in flat three-dimensional space. In chapter 1 the surface of the sphere was taken as a useful illustration of this. The rules of Euclidean geometry do not apply to the surface of the sphere, as may be readily appreciated by inspection of fig. 4.9. To understand this figure, first note that it is not possible to draw *straight* lines on a curved surface. However, between any two points in the surface we can draw a *straightest* line in the

Fig. 4.9. Curved space. The surface of the sphere is a curved (mathematical) space. Clearly the angles of a triangle do not add to make two right angles in this space. Nor do the circumferences of circles always increase with their radii. In a small enough region, the space is approximately flat, and Euclid's geometry applies.

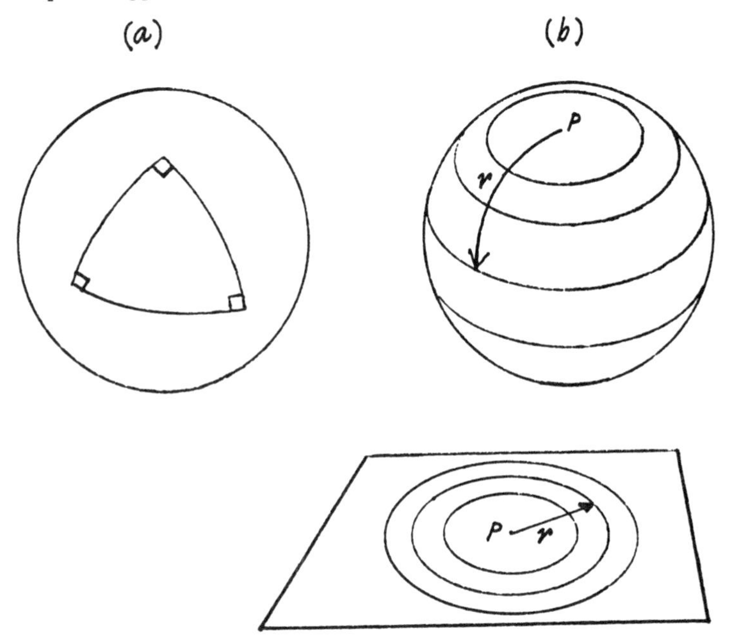

sense of the shortest path between the two points. This shortest path on a curved surface is known as a *geodesic*. On a spherical surface geodesics are segments of great circles, which mark the intersection between the surface and a plane passing through the *centre* of the sphere. On the surface of the Earth, aeroplanes usually fly close to geodesics to minimise the distance of travel. A geodesic from New York to Tokyo passes very close to the North pole, so that the pilot has to fly first north and then south again to achieve a 'straightest' path.

The properties of geodesics depend on the nature of the surface on which they are drawn. On a flat surface it is always possible to find *parallel* geodesics. This is not possible on a spherical surface, because any two great circles will intersect twice (for example, the meridians – lines of longitude – on the Earth all intersect at the poles even though they appear parallel at the equator).

In fig. 4.9*a* a spherical triangle has been drawn, the sides of which are geodesics. It is immediately seen that the interior angles of this triangle add up to *three* right angles (270°) rather than two as in the case of flat-plane geometry.

Another example which is illustrated in fig. 4.9*b* shows concentric circles drawn around a point P on the surface of the sphere, and also on a flat surface. As is well known, there is a constant ratio (2π) between the circumference of a circle and its radius, on the flat sheet. However, on the sphere, the circumference of a circle of a given radius is clearly smaller than $2\pi r$, and actually *decreases* with increasing r once r has extended more than a quarter of the way round the surface. That is, there exists a *maximum* circumference for these circles. Notice, however, that in both examples given here, for a sufficiently small region of the sphere, Euclidean geometry remains a good approximation. On a small scale, the surface is nearly flat.

Although the geometrical features of the curved surface have here been illustrated by embedding the surface in flat three-dimensional space, a two-dimensional individual entirely restricted to the surface of the sphere, and without access to our own three-dimensional 'overview', could nevertheless deduce from observations made entirely *within* the surface that, for example, the angles of large triangles do *not* add up to two right angles. He could also deduce that it was impossible to draw parallel lines in his space, and many other geometrical properties characteristic of the sphere as well. One way of expressing this fact is that these properties are not simply features of the way we have chosen to embed the surface in the surrounding space; they are *intrinsic* to the surface itself. In addition to the intrinsic geometrical structure, an ambitious two-dimensional inhabitant of the surface could also deduce from distant travel and observation *topological* properties of the sphere; for example, the fact that it is a closed, finite surface.

Although the considerations so far have been restricted to two-dimensional spaces (i.e. surfaces), all the general results can be carried over to spaces of higher dimensions. There clearly exists the possibility that real three-dimensional physical space (and even four-dimensional space–time) obeys the rules of, say, spherical geometry rather than Euclidean geometry. The whole of space might possess an intrinsic geometry which is non-Euclidean.

A curved three-dimensional space possesses certain peculiar properties. For example, suppose three-dimensional space were analogous to the sphere in two dimensions. In the latter case there is no constant ratio between the radius and circumference of circles, and a maximum-circumference circle exists. In the three-dimensional case, the circles are replaced by spheres and their circumferences replaced by the surface areas of the spheres. It is well known that the rules of Euclidean geometry require a constant ratio (4π) between the square of the radius r and the surface area of these spheres. In the same fashion as the two-dimensional analogy, a spherical three-dimensional space has the property that these surface areas are in general *smaller* than $4\pi r^2$. Moreover, for sufficiently large spheres there is a *maximum* surface area, after which increasing the radius of the sphere actually *decreases* its total surface! The total volume of such a space is finite. A property as peculiar as this is quite amenable (in principle) to observation in the real world. In chapter 5 it will be discussed what evidence there is that the large-scale geometry of the universe might be spherical.

Einstein's daring and revolutionary proposal was to link together these mathematical ideas of curved geometry with the physical properties of gravity. He suggested that space–time in the presence of gravity is not 'flat', obeying the rules of Euclidean geometry, but bent into a more complicated geometrical structure. A freely-falling particle then moves through this curved structure along a straightest possible path – a geodesic. If gravity could be absent, space–time would be flat. Then this particle path would reduce to the familiar Newtonian force-free situation – uniform motion in a straight line. According to these new ideas, a reference frame which is falling freely at one location in an inhomogeneous (i.e. non-uniform) gravitational field will register a distortion of space–time around it as the bent geometry causes the space–time to curve away. Particles falling freely at distant points therefore follow curved paths, exactly in accordance with the experiences outlined on page 91 for the observer in the falling box.

In chapter 1, it was mentioned that Newton had discovered his mechanics by enquiring into why bodies accelerate, not why they move uniformly. He regarded inertial motion as natural and not in need of explanation. Forces were needed to *change* the uniform velocity of a body, not perpetuate it. Gravity was considered to be such a force, causing a body to fall ever faster towards the ground. Then, in the twentieth century, Einstein progressed a step further. He regarded a *falling* body as being in a natural state of motion, but in a bent space–time. There is nothing mysterious about gravity. What must be explained is not why the apple falls, but why it stops when it hits the ground. Just as Newton abolished forces for uniform motion, so Einstein abolished forces for falling motion. Only deviations from free fall require forces. When the apple stops, violent non-gravitational forces throw its space–time path away from the natural geodesic. In contrast, the Earth orbits in a curved path around the sun not because it is *forced* from straight-line motion, but because it drifts unhindered through the curved space–time in the sun's vicinity. This brilliant identification of gravity with geometry is hailed as one of the greatest triumphs of human intellectual activity in history.

Of course, a descriptive account of gravitation in terms of geometry as given here does not itself constitute a physical theory. Einstein had to provide a set of mathematical equations which describe precisely *how* a given source of gravitation curves the space–time manifold in its vicinity. In his search for the correct equations he was guided by several fundamental principles. For example, in the limit of weak gravitational fields and small velocities, the new theory should reduce to Newton's theory of gravitation. This requirement is essential because the Newtonian gravitational model has been (and still is) used with such spectacular success for generations. Secondly, the general theory of relativity must reduce to the special theory for weak gravitational fields.

In Newton's theory, the *mass* of a body acts as the source of its gravity. This quantity is not a suitable source in a relativistic theory, because mass is equivalent to energy (through $E = mc^2$), which in turn is coupled to momentum, in much the same way that space and time are coupled in relativity. Furthermore, momentum is intimately related to mechanical stress and pressure. Consequently, any new theory of gravity which is consistent with relativity ought to allow for all these physical quantities – stress, energy and momentum – to generate gravity.

The next step is to find the correct space–time geometrical quantity to couple to this source. By examining how stress, energy and momentum are related, Einstein was able to find geometrical quantities describing the space–time curvature which are related together in exactly the same way. By equating these two quantities together – one geometrical, one physical – he arrived at his famous *field equations*. These equations describe in detail how a particular distribution of stress–energy–momentum will distort the space–time fabric in its vicinity.

One of the most notorious features of Einstein's field equations is the extreme difficulty in solving them. Indeed, in the 60 years since their discovery, only a handful of exact solutions are known. However, even as early as 1916 someone had found one of the simplest, and still the most important, exact solution, corresponding to the empty space–time in the vicinity of a spherical body. This solution was found by Karl Schwarzschild

(German astronomer, 1873–1916), and is known, appropriately enough, as Schwarzschild's solution. In spite of its simplicity – a spherical mass surrounded by empty space – this system provides an excellent model of the solar system, with the central body representing the sun, and the empty vicinity being the region in which the planets move (the gravity of the planets themselves is neglected). By calculating the geodesic paths in the Schwarzschild space–time, one obtains the shape of the planetary orbits. This problem had long ago been tackled by Johannes Kepler (German, 1571–1630) and Newton using Newtonian theory. The outcome of their earlier computation was the prediction that the planetary orbits were ellipses with the sun at one focus, a monumental result known to be in good agreement with observation.

The general theory of relativity also gives a result very similar to that of Newton's theory in the case of our solar system, but there are slight, and crucially significant, differences. Instead of predicting an exactly elliptical motion, Einstein's theory describes an elliptical path which precesses, or rotates its axes, in the same plane (see fig. 4.10). The effect is very small. For the planet nearest to the sun, Mercury, it is greatest, but the precession still only amounts to 43 seconds of arc per century; that is, one complete rotation of the ellipse every three million years. To make matters worse, several other effects also cause a precession of Mercury's orbit, and moreover, these effects are much

Fig. 4.10. Curved space-time twists Mercury's orbit. Johannes Kepler discovered the motions of the planets to be ellipses. Newton explained this using his theory of gravity. Einstein's theory also predicts an elliptical path, but one which rotates slowly in its own plane. For Mercury this turning motion is a mere 43 seconds of arc per century.

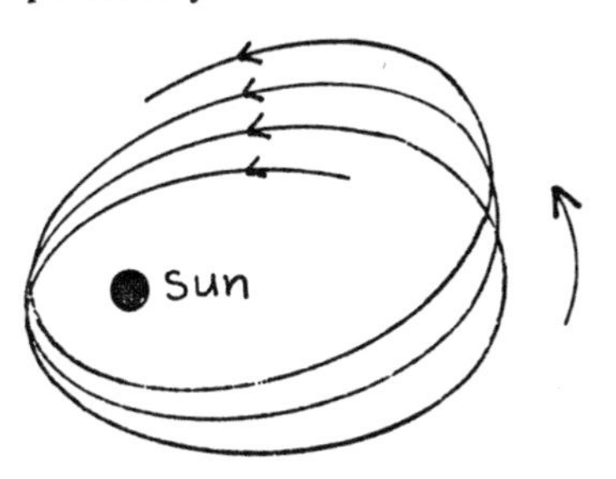

larger. Nevertheless, they can be calculated and allowed for, and indeed, it was known before Einstein published his theory that there was a discrepancy of about 40 seconds of arc per century in Mercury's orbital precession. The brilliant and exotic explanation of this discrepancy by Einstein in terms of space–time curvature is one of the very limited number of observational confirmations of the general theory of relativity.

So much for the effect of space–time curvature on space, and on the paths of test particles moving in space–time. Let us now turn our attention to the effect on time, and the way in which the distortion of space–time modifies the rates of clocks placed in gravitational fields.

One of the features which many theories of gravity, including the general theory of relativity, possess is that clocks in a strong gravitational field, such as that near the surface of a large spherical mass, run more slowly relative to an identical clock

Fig. 4.11. The Schwarzschild radius. A photon of light has a quantum mechanical frequency proportional to its energy E. According to special relativity this energy also has an associated mass mc^2. The Newtonian gravitational force on mc^2 due to the mass M (concentrated at a point for simplicity) is GMm/r^2, so the energy lost when the photon climbs away from a distance r to a very distant point is GMm/r. This represents a fractional energy loss, and hence fractional frequency loss, of GM/rc^2. Using the frequency of the light as a clock, this represents a proportional slowing down of the clock rate at r, as seen by a distant observer. This time dilation will become very large, and the intensity of the escaping light very dim, when r is comparable with GM/c^2, for the fractional energy loss then approaches unity. The theory of relativity actually requires this situation to occur when $r = 2GM/c^2$. This distance is known as the Schwarzschild radius, and is denoted by r_s in the figure. Most objects in the universe are much larger than their Schwarzschild radii, and for these the time dilation effects are barely noticeable.

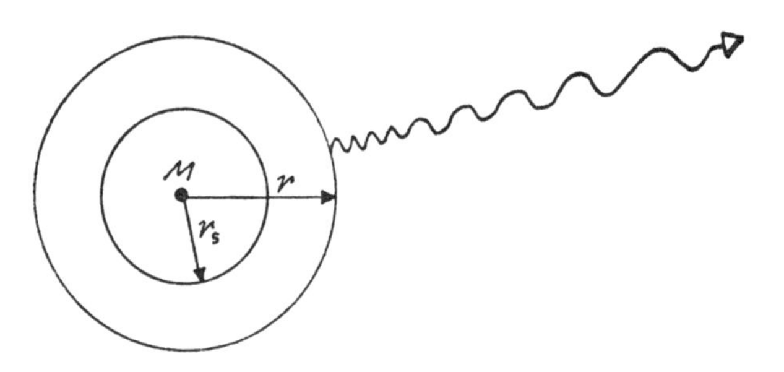

placed a long way away from that mass. This gravitational time dilation is naturally also embodied in Schwarzschild's solution, the details of which are beyond the scope of this book. However, the reader who has a passing acquaintance with the quantum theory will find a heuristic treatment of this effect given beneath fig. 4.11. Once again, the effect is very small in the solar system. The fractional time dilation at the surface of the Earth is a mere ten-millionth per vertical centimetre. Nevertheless, even this minute effect can be measured using a type of atomic clock. Of course, what is being discussed here are *relative* clock rates. The reader should not imagine that time would seem to pass more slowly near the Earth's surface than out in space. It is only that clocks in these different locations would get gradually out of synchronisation.

4.3 Black holes: space–time in collapse

If the effect of gravitation on space–time were limited to the minute effects discussed above, the general theory of relativity would remain a rather isolated branch of physics; a sort of intellectual ornament. In recent years, however, it has become apparent that there may well exist in the universe fantastic objects whose gravitation is so strong that they modify the

Fig. 4.12. Curvature of the line is measured by the radius of the touching circle. When the radius is *small*, the curvature is *large*.

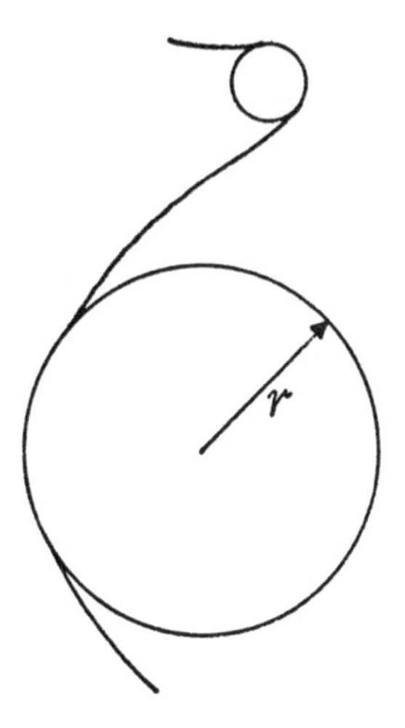

properties of space and time in their vicinity in strange and fascinating ways.

It is worth examining the use of the word 'strong' here. When is a gravitational field to be considered strong? In this context, when the curvature of space–time is large. To get some idea of the circumstances under which this will happen it is necessary to appreciate how curvature is measured. In fig. 4.12 a curved line is drawn. At two points the local curvature of the line is approximated by touching circles. The radius of these circles may be used as a measure of the curvature of the line; the smaller the radius the larger the curvature. A similar procedure may be adopted for the curvature of space–time. At each point a measure of curvature may be defined. An indication of gravitational strength can then be obtained by comparing this measure with ordinary distance units.

Suppose for a moment that all the gravitating mass in Schwarzschild's spherically symmetric solution were concentrated at a point. At a great distance from the mass the curvature of space–time is small, but as the mass is approached the gravitational field rises and the curvature of space–time becomes very large. At a certain distance from the point mass, the curvature will be comparable with this distance itself, and the corresponding space–time distortion will be very large. This critical distance must depend on the mass of the object, and on the gravitational constant, G, and we could also include the speed of light, c, because this connects the units of space and time, as explained in chapter 2. The only way in which G, M and c can be combined together to give a quantity with the units of distance is as GM/c^2, so that apart from a numerical factor this quantity must determine the radius from the point mass at which the gravitational distortion of space–time is very considerable. The same result was deduced on quantum grounds in connection with fig. 4.11. The radius GM/c^2 was also discovered as long ago as 1796 by Pierre Laplace (French, 1749–1827) as (one half of) the distance inside which the Newtonian *escape velocity* from a mass exceeds the velocity of light.

Schwarzschild's solution gives this critical radius exactly, as $2GM/c^2$, so this distance is now known as the Schwarzschild radius. Putting in some numbers reveals that the Schwarzschild radius for the Earth is about a centimetre and for the sun about a kilometre. In both cases therefore, these objects are enormously greater than their Schwarzschild radii, in accordance with the experience that the space–time distortion near them is exceedingly small.

It must not be imagined that this distortion becomes large near the centres of these objects. Schwarzschild's solution applies just to the *exterior* region – the empty space outside the central mass. Only if the entire object were shrunk down to near its Schwarzschild radius would the effects of curvature become large. Then, a typical star like the sun, being shrunk to a mere kilometre or so in diameter, would be at a colossal density – about 10^{25} times that of water! For the Earth the density would be about a million times higher still, with its entire volume shrunk to less than the size of an egg! At the other extreme, a galactic mass need only be about the density of water, while the critical density of the entire universe is only about a hundred times the actually observed density of luminous matter.

One of the properties of an object which has shrunk to near its Schwarzschild radius is that light leaving its surface will lose nearly all its energy in escaping the tremendous pull of gravity. Consequently, the surface of such an object will appear to a distant observer to be very dark. This was the basis of Pierre Laplace's 1796 conjecture, purely on the basis of Newtonian gravitational arguments, that there might be massive objects in the universe that were completely black because no light energy would be able to escape their enormous gravity. The time dilation at the surface of a highly compact object, observed in this dwindling luminosity, will be almost infinite. Events there will appear to take place so slowly that the surface will seem almost 'frozen' still. For this reason such hypothetical objects were once referred to as 'frozen stars' though the name is somewhat misleading because in practice their surfaces would appear to be almost totally black. A more appropriate terminology now in popular usage is *black hole.*

Theoretical astronomers have constructed a number of scenarios in which a black hole might be formed in the real universe. For example, it is widely accepted that about 10 billion years ago the universe was in a very dense condition, with the now highly dispersed matter enormously compressed. Under these circumstances, local condensations of matter could have become trapped by their own gravity into microscopic sized black holes, no larger than a subatomic particle, but with a mass of about 10^{15} grams. At the other extreme, objects with masses equivalent to millions of stars might find it reasonably easy to collapse to the critical black-hole radius because the corresponding required density is in this case unexceptional.

Perhaps the most convincing scenario for black-hole formation occurs at ordinary stellar masses. In recent years it has become widely believed that the black hole represents the natural end point of the life of some massive stars. To understand why this should be so, it is necessary to take a short digression into the structure of stars.

Most stars are similar to our sun, consisting mainly of the lightest substance, hydrogen, and measuring about one-and-a-half million kilometres in diameter. They are not particularly dense, though, for the following reason. Gravity of the stellar material tries to pull the hydrogen towards the centre, but in so doing the contracting gas gets hot. Near the centre it is so hot (several millions of degrees) that nuclear fusion – the power behind the hydrogen bomb – occurs. The fusion process consists of a competition between the short-range attractive nuclear forces (based on the strong interaction) between the protons and neutrons in the atomic nucleus, and the long-range electric repulsion between the charged protons. In the centres of stars, the temperature is so hot that the rapidly moving atomic nuclei can collide with sufficient violence to overcome their electric repulsion and come close enough for the much stronger short-range nuclear attraction to operate. This causes the light nuclei (for example, hydrogen) to fuse together into heavier nuclei (such as helium). In doing so, part of the total mass is converted to energy in the form of gamma ray photons and neutrinos.

This energy has two effects. Firstly, it keeps the star hot so that the fusion continues; in fact, stars like our sun are in a stable, steady-state condition, with their heat loss to the surrounding space just balanced by the fusion energy production. Secondly, the pressure from this energy prevents the outer layers of the star from collapsing inwards, thus ensuring that the density of stellar material is rather low (less than water near the edge of the sun). The fusion process which produces most of the energy is the conversion of hydrogen into helium, the next lightest element. It follows that eventually the hydrogen will become rather scarce, and the balance which maintains the stability of the star will begin to fail. The star will then enter on to a career of rather violent activity. This is not expected to occur to our sun for a matter of thousands of millions of years, but for stars a few times more massive than the sun (not unusual) it has already happened. The precise details of the stages that follow need not concern us, but the death of a star must be either a catastrophic explosion or gravitational collapse. It has long been known that when the nuclear fuel of a star is exhausted, collapse to a very high density cannot be prevented. Some very dense stars are actually observed – they are called white dwarfs, and their surface gravity is thousands of times greater than that of the sun. The white dwarf substance is so dense that one ton of ordinary matter would be squashed up into the size of a thimble.

Further collapse of white dwarfs is prevented on account of a rather subtle quantum mechanical effect known as electron degeneracy pressure. However, this new effect cannot support a mass greater than 1.44 times that of the sun. Still denser objects than white dwarfs can occur, in which the gravitational force of the star is so strong that it actually crushes the individual atoms into neutrons. These so-called *neutron stars* are incredibly dense; they are only a few kilometres across and yet possess masses comparable to the sun. A mere teaspoonful of neutronic matter on the surface of such a star would weigh 100 million tons. If the ocean liner QE2 were dropped on to a neutron star it would be crushed up to the size of a grain of rice!

Neutron stars are held apart by neutron degeneracy pressure, but once again, only reasonably light stars can be so supported. For a really massive star there appears to be nothing else available to support its material against total collapse.

Although the precise details of the collapse depend to some extent on the properties assumed for the dense neutronic matter, the general features are well understood. In dealing with the gravitational collapse of objects beyond the neutron star stage we are in the region of intense gravitational fields where Einstein's general theory of relativity must be used to describe the system. Only simplified models are amenable to direct mathematical analysis by this beautiful but technically complex theory. For example, the collapse of exactly spherical objects, whose exterior geometry is described by Schwarzschild's solution, has been extensively studied. A good first approximation is to assume that all parts of the star fall inwards freely, i.e. to neglect the internal pressure.

From the viewpoint of an observer fixed to the surface of such an object events proceed at an alarming pace. The 'star' implodes so rapidly that its surface shrinks to a fraction of its size in much less than a second. Scarcely has the observer had time to blink than he is swept through the critical Schwarzschild radius. Being in free fall he is weightless, and passes through this critical surface without noticing any unusual effects on the space–time in his vicinity. However, this sequence of events as witnessed by a very distant observer (who does not fall inwards with the collapse) is very different indeed. As the radius of the star nears the Schwarzschild radius, its surface approaches the region of infinite time dilation. These stages of the contraction therefore appear from afar to be progressively slowed down until the time scale is so extended that no further collapse is perceptible. The object simply hangs suspended outside its Schwarzschild radius, frozen in time for ever.

If no further change takes place, a distant observer cannot expect to see the star when it has collapsed *inside* its Schwarzschild surface, however long he waits. The events which occur to the star after its passage through this surface are totally inaccessible to the outside universe (being 'beyond infinity' in time). The Schwarzschild radius therefore plays the role of an *event horizon*, separating those events on the star which can be seen by the outside world from those that cannot. Nothing that occurs inside the horizon can influence or affect the outside in any way whatsoever.

The qualitative features of the Schwarzschild black hole are easy to understand by looking at a diagram of light waves in its vicinity.

The effect of gravity on light can be conveniently represented by dots and circles shown in fig. 4.13. Here the dot is supposed to be a point in space where a little burst of light is emitted in all directions. The circles are the positions of the wavefront at successively later moments. In the absence of gravity (*a*), the wavefront spreads out uniformly in all directions. However, as

Fig. 4.13. One manifestation of a distorted space–time is that light appears to be dragged about by gravity. In the diagrams the black dot denotes a point in space at which a flash of light is emitted in all directions. The circles represent the positions of the spreading wavefront at successively later moments. In (*a*) there is no gravity, and the waves spread out equally in all directions. In (*b*), a gravitational field acting to the left displaces the circles somewhat in that direction. Diagram (*c*) is an enlargement of the situation at the Schwarzschild radius shown in fig. 4.14. The dot is at a *fixed* distance from the centre of the star, and the circle is displaced so much that the right-hand edge, which represents light directed *away* from the star, cannot make any progress at all. The dot will never be seen by a distant observer (to the right) however long he waits. Diagram (*d*) shows the situation inside the Schwarzschild radius. The dot is again at fixed distance from the centre of the star. The displacement of the light circles is now so great that the wave surfaces directed towards, *and* away, from the star *both* move in towards the star, trapping any material body inside them as they go.

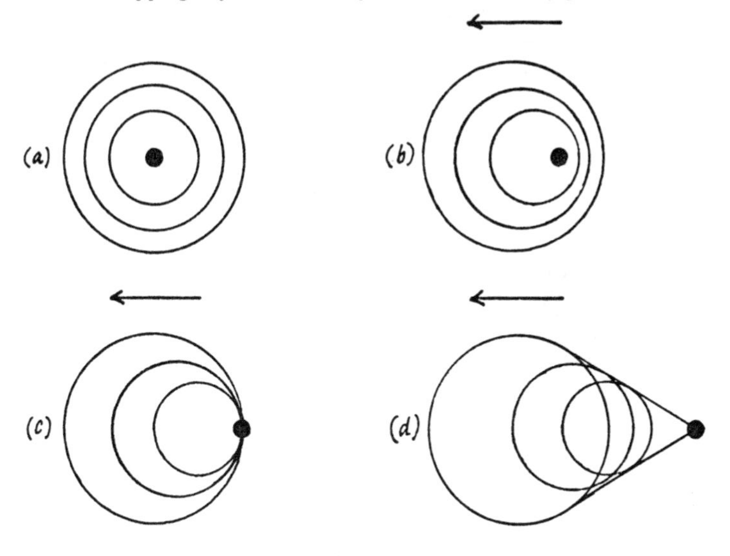

depicted in (*b*) the effect of a gravitational field is to drag the wavefront to one side. Of course, this distortion is a manifestation of the bending of space–time already discussed.

In fig. 4.14 the situation near a black hole is described. The central disk is the star, shrunk well inside its Schwarzschild radius which is marked with a broken line. At a great distance from the object, the wavefronts (only one circle is shown for simplicity) are hardly distorted. As the event horizon is approached, the light is dragged over more and more strongly towards the star. The horizon itself is distinguished by the circumstance that the side of the wavefront remote from the star cannot progress away from the star *at all.* This is denoted by drawing the dot on the edge of the circle of light (an enlarged version is shown in fig. 4.13*c*). The wave is still expanding locally at the speed of light, but here the dragging is so intense that the best the light can do is to mark time and remain stationary relative to the star. It follows that however long an observer far away from the star waits, he will never see this struggling wavefront.

The situation is rather like a runner on a moving track. As

Fig. 4.14. Behaviour of light waves in the vicinity of a black hole. Inside the Schwarzschild radius (broken circle) even the outward directed light waves move in towards the centre of the star. Events in this region of space–time will therefore never be seen by a distant observer. He will only see light waves from the region outside the broken circle. The interior region is both black and empty – a black hole.

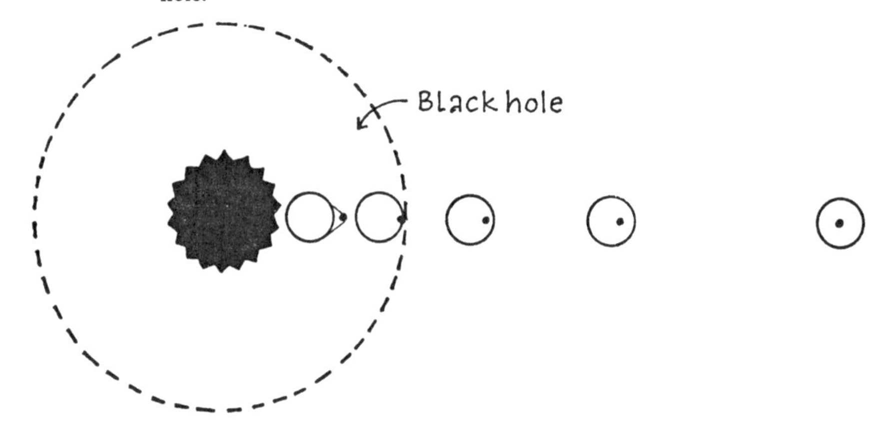

fast as he tries to run along the track, the track itself pulls him back. Here the runner is the light beam, and the track is space–time which, in a very crude sense, is collapsing into the black hole.

Inside the event horizon, the wavefronts are so greatly displaced inwards (i.e. towards the centre of the star) that they have moved entirely beyond the dots (see fig. 4.13*d* for an enlargement). Physically this means that both inward and outward directed pulses of light, although still separating from each other at the speed of light, are actually both moving inwards, towards the star. It follows that the event which produced the light can never be seen outside the black hole. For these light waves there is no escape. The farther inwards they are pulled, the greater the dragging becomes. In a very short while they are drawn right into the centre.

Fig. 4.15. Inside the black hole! Once the observer ventures across the Schwarzschild radius he cannot get out again, however hard he tries. Even with rockets blazing, he gets dragged inexorably downwards by the collapsing space–time geometry. Light signals sent towards the outside also get sucked back. There can be no escape, for, in the incredibly brief moment that he takes to fall into the black hole, all of eternity has passed on the outside.

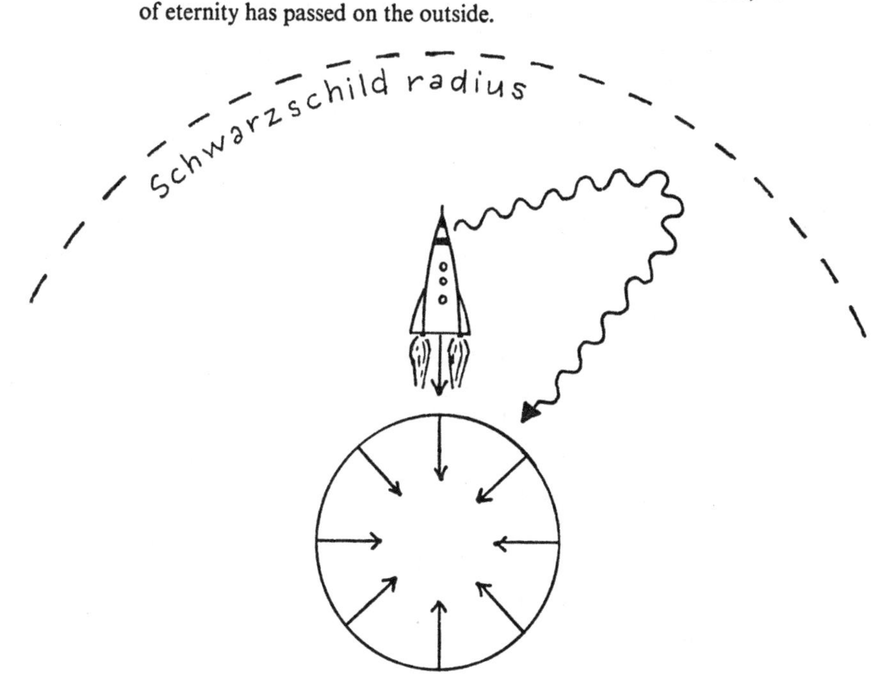

Because any material body cannot exceed the speed of light, it is clear that an observer who finds himself inside one of the circles of light must remain inside the circle as it is dragged inwards. Consequently, inside the event horizon, he too will be drawn irresistibly towards the centre of the star. Even with the most powerful rocket in the universe, it is no longer possible for him to escape to the outside universe again or even, for that matter, to remain stationary. No force can prevent him from plunging on inwards. Nothing can get out of this black hole region. He is trapped for ever in the most secure prison in the universe, unable even to communicate his fate to the outside world. The star itself, being made of matter, must share in this fate of being dragged inexorably inwards on itself.

Although the details of the collapse phase depend to a minor extent on the internal constitution and structure of the star, its frozen end state as viewed from a distance is quite independent of this. Because of the gravitational red shift, the surface luminosity fades away exponentially fast, and after a mere few thousandths of a second nothing can be seen of the imploding surface except blackness – a black hole. Consequently, it is not possible to tell by looking at the black hole what it is made of; any two Schwarzschild black holes of the same mass look identical.

More than this, they cannot even be distinguished from each other by measuring subtle physical effects, such as nuclear forces or magnetic fields. According to present theory, whether the black hole is made from ordinary matter, anti-matter or even from neutrinos, it would be impossible to tell the difference by any known physical means.

It is fascinating to understand how the black hole prevents the external observer from knowing what it is made of. It is no use trying to lower a rope down into it with a grab on the end to claw up a sample from the surface, for the intense gravity acting on the lower end of the fixed rope would tear it apart however tough the material from which it was made. A lightning rocket trip to the contracting surface and out again would be equally futile. For when the rocket landed, the surface would no longer appear static, as it does when viewed from a great distance. The landed rocket would be in the falling reference frame of the surface, and would be swept across the horizon along with the star. Furthermore, beyond a certain moment after the formation of the black hole, the rocket could not even reach the surface before passing inside the horizon, so it would have to pursue the retreating matter across the Schwarzschild radius quite unable to return to the outside world with the sample thereby obtained.

Nor would it be possible to illuminate the surface by shining a strong light into the black hole. The light would suffer the same fate as the rocket, and pass across the horizon, its reflection from the imploding matter caught by the collapse and snatched back downwards by the overwhelming gravitation.

All attempts to stick 'labels' on the black hole would be equally fruitless. For example, a pattern of electric charges placed on the collapsing matter need not disappear as it approaches the Schwarzschild radius, but the electric field it produces in the exterior region gets warped and distorted by the curved space around the object, so that all of the charge appears from a great distance to be concentrated at the centre of the matter. Hence it is not possible to 'read' the structure of the pattern any more. Calculations have been performed to test the 'readability' of all kinds of 'labels', using strong and weak, as well as electromagnetic, forces. In all cases the conjecture has been confirmed that only the total mass and total electric charge leave their stamp on the exterior physics. All other structure and information gets washed out in the collapse.

These considerations have also been extended to the case of non-spherical collapse. Calculations indicate that if an object possesses 'bumps' then as it collapses these bumps do not disappear, they actually grow. However, it is no use labelling a black hole with a pattern of bumps, because once again the curved space around the black hole makes the pattern unreadable from afar. To a distant observer the resulting black hole would still appear spherical.

It is now possible to appreciate the appellation 'hole' in this connection. Although the collapsing matter appears from a distance to be permanently suspended over the Schwarzschild radius, the nature of this matter is apparently in no way discernible. Its structure cannot affect physically the outside distant region, so that to all extents and purposes the matter is inaccessible – it has retreated out of the universe. The dark sphere which remains is like a hole in space – just a black hole.

So far most of the discussion of gravitational collapse has concentrated on the appearance of the collapsing matter from a great distance. Weird though this view of events may be, the experience of the observer fixed to the surface of the collapsing object is positively bizarre.

As he falls down with the collapsing matter the surface gravity of the object progressively rises. The increasing gravity has no effect locally on a freely-falling particle, but an extended body, such as a real observer, would begin to feel *tidal* forces for the following reason. The collapsing star is becoming so compact that the strength and direction of the gravitational force is varying even over the dimensions of a human body. If he fell feet first, the observer would be stretched, because his feet would be nearer to the gravitating star than his head. In addition he would be crushed, as the diminishing volume of space around him tried to squash him up. Let us assume that the observer is a very small, very strong individual who can survive for quite a while the rigours of these tidal forces. As he falls on downwards, the landscape around him continues to possess complex structure, perhaps even regions of electric charge – all of the things that are becoming invisible to the outside world remain accessible to the falling man.

This scenario is hardly changed at all by including the effects of pressure. For matter under ordinary conditions, by making it stiff enough, it can resist being crushed. However, recall from chapter 2 that matter moving at near light velocity cannot be totally rigid because of the limitation of all physical influences to sub-luminal speeds. Thus, matter which is totally incompressible when at rest, necessarily becomes squashy when falling rapidly down a black hole. In a black hole, however *stiff* the matter is, the crushing continues. Paradoxically, because pressure is also a source of gravitation in general relativity, a resisting body is actually forced inwards more strongly. Whatever efforts are made to resist falling inwards, all matter (including the observer and his rocket) must reach the centre of a spherical black hole of stellar mass no later than about a ten-thousandth of a second (measured time) after passing through the horizon.

This conclusion raises one of the most enigmatic questions in modern science. What happens at the centre of a black hole when the matter finally arrives there? Sometimes the question is waved aside with the remark that whatever occurs inside a black hole can never be of any consequence to the universe outside anyway. But such remarks do not discourage scientific curiosity and attempts to answer this fascinating puzzle have frequently been made.

Before considering these attempts, the reader will do well to place the discussion in perspective. The general theory of relativity, on which all studies of black holes depend, is only a *theory*. Its predictions have been reasonably well checked only in gravitational fields in the solar system. Inside a black hole, gravity is billions of times stronger than this. No one knows how far the theory can be extended with any confidence, or which of its features might remain if a better theory were known. The theory of relativity is very beautiful and accepted by most physicists as the best description of gravity available. But all theories have their limits. It is fun to see how far general relativity will take us, and we may discover something important on the way. It is the only way we can try to find out what happens inside a black hole without actually falling in one. Nevertheless, it should not be forgotten that we are talking about the world of model black holes, not the real world.

As the spherical implosion proceeds inexorably, the density of matter inside the star rises faster and faster. It is already incomprehensibly compressed, its properties are entirely unknown. The tidal forces and space curvature of the surface of the collapsing matter also rise ever more rapidly, eventually smashing all possible structures. The theory of relativity definitely predicts that, unless something very strange happens to the matter, there is no point at which this implosion can be halted. As always when we extrapolate a physical theory to its absolute limit we reach an absurdity. In this case the absurdity is the prediction that all the matter in the star is squeezed into the same mathematical point. At this point the density and the space curvature must be infinite. This absurdity is called a *singularity*. A singularity is not an object, it is a place where known physics has come to an end.

At one time it was conjectured that the singularity was a consequence of the simple nature of the model used for studying gravitational collapse. After all, if we require the imploding star to be always precisely spherical it has no alternative than to shrink to a point if the collapse cannot be arrested. It seemed reasonable to suppose that in a real system the matter might fall very close to the exact centre, but not precisely through it. Remarkably, the appearance of a singularity does not seem to depend crucially on the model of the collapsing matter after all. In a series of theorems which deal only with the *topological* structure of space–time, two British mathematicians Stephen Hawking and Roger Penrose have proved that, as long as the energy and pressure of the collapsing matter do not undergo some very implausible changes, a singularity of some sort is then inevitable.

Unfortunately, the theorems say very little about the nature of the singularity which forms. The concept of singularity which these theorems employ is much less precise than that which occurs at the centre of a spherical implosion. All that can be said is that some particle path through the space–time must come to an end. That is, a particle which is actually dropped along this path could not remain in space–time. Sometimes this is expressed by saying that space–time develops an edge or boundary, or that it comes to an end at the singularity. Occasionally it is conjectured that any matter which encounters a singularity leaves space–time altogether. The Hawking–Penrose theorems do not say whether the collapsing matter will actually strike the singularity or not. However, if the matter does not leave space–time through this route, it certainly cannot come back into our universe again, for it is trapped by the event horizon. There are some model black holes, more general than Schwarzschild's solution, that allow for the possibility of the hole to rotate and carry electric charge. Some of these idealised special cases display the curious feature that matter may actually avoid the singularity and pass into new regions of space–time, which are joined on to ours through the inside of the black hole. These 'other universes' are, of course, hidden behind the event horizon. Such special models are useful in telling us something about the nature of space–time. They should not be considered as models of the real universe.

There is no doubt that the prediction of singularities in space–time has profoundly disturbing consequences. Because we cannot continue physics into such a region, we cannot predict what might come out of it. What happens in a part of space–time in the causal future of a singularity is totally unpredictable.

This unpalatable feature of singularities has led Roger Penrose to propose a 'cosmic censorship' hypothesis. This hypothesis forbids the formation of singularities unless they are contained inside event horizons. A singularity inside a black hole cannot invade the universe outside with unknown influences. Mathematical studies have shown that cosmic censorship is very likely to be correct, but there is still no proof available.

If cosmic censorship fails, even in principle, then it is possible that 'naked' singularities (visible from a great distance) could form. Some writers have dwelt morosely on the disastrous consequences of such a possibility for the rest of the universe. However, it should always be remembered that present theory can only be an approximation. For example, inside the incredibly small distance of 10^{-33} centimetre, quantum effects (to be briefly discussed in sections 4.4 and 4.5) will modify gravity in a fashion which present theory is as yet quite unable to describe. Nobody knows what quantum gravity will have to say about singularities. In another few years, theorists may have a completely different picture of what a singularity is all about.

4.4 The black hole as a powerhouse

The Schwarzschild black hole is not the only kind of black hole possible. Other types are known to theorists. These could in principle be formed out of electrically charged or spinning matter. The general pattern of collapse, horizon and singularity is the same for these other types of black hole.

However, some fascinating new features appear when their space–time structure is probed using mathematics.

The effects of rotation on the properties of the space–time around a spinning black hole are depicted in fig. 4.16. This is a section through the centre of the black hole perpendicular to its rotation axis. Once more, wavefronts of light are dragged inwards towards the centre. But in addition the rotation of the object has the effect of dragging the wavefronts around with it. Consequently, there is both an inwards and transverse displacement of the circles drawn in the figure. Inside a certain distance from the centre, called the static limit, these circles are displaced entirely beyond the dots. In this region an observer, who must remain inside the circles, will find it impossible to remain still relative to a distant observer, who is far from the black hole. Picturesquely speaking, he is caught up in a kind of space vortex. Any particle of matter which falls into this vortex gets swept around in an invisible whirlpool of space. No amount of rocket power, no force in the universe, can prevent the particle from being dragged around; indeed, dragged faster than light relative to the distant observer.

Fig. 4.16. Rotating black hole. Light wavefronts are dragged both inwards *and* around. Inside the outer circle (static limit) the dragging is so strong that the light waves directed with, *and* against, the rotation direction *both* move anticlockwise. Light signals from the wavefront edges remote from the centre of the star can still escape to a distant observer from this region. Further inwards, inside the broken circle, the drag towards the centre is so great that even the outward directed edges of the wave front are moving inwards. This is the black hole region of space–time from which no light can escape.

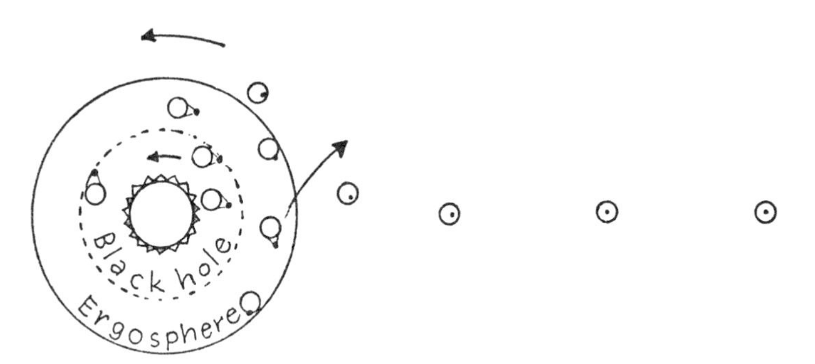

In spite of being caught in this irresistible rotator, it is still possible for someone who ventures into this region to get out again and return to a safe distance. The reason for this is apparent from fig. 4.16. Although the circles are displaced completely beyond the dots, the *direction* of the displacement is not directly towards the centre of the system, but to one side. The edge of the wavefront remote from the centre still moves gradually outwards.

Some of the light waves can thus escape, and hence, so too can an observer travelling slower than light. Nearer the centre there is a true event horizon. Inside this inner surface the radial dragging is great enough to prevent any light or matter from escaping.

The strange region between the static limit and the event horizon is called the *ergosphere*, because of the following curious possibility, discovered by Roger Penrose. If, during its sojourn in this region, a body breaks into two pieces in such a way that one of the pieces falls down into the black hole, then according to calculations, if the circumstances are right, the remaining piece emerges again with more energy than it went in with! What happens is that some of the rotational energy of the black hole gets transferred to the emerging particle. This is the reason why this region is called the ergosphere, after the Greek word *ergos*, meaning work. In principle, it is possible to enhance the energy of the emerging particle by an amount equal to the entire rest mass of the part falling inwards, making the rotating black hole the most efficient energy conversion machine known. This 100% conversion efficiency should be compared with the mere 1% efficiency of nuclear fusion which powers the sun.

The effect of this peripheral manipulation on the black hole itself is to slow down the rotation somewhat. The energy supply by the Penrose process is thus limited. Nevertheless, in a fantastic story related in a recent analysis of this process, a description has been given of an imaginary civilisation who take up residence around a rotating black hole. Each day their rubbish and technological pollutants are taken down into the ergosphere in trucks, there to be deposited through the horizon, whilst the trucks return carrying the energy equivalent of this waste to power the community's energy supply. The rotating black hole is the ultimate efficient dual-purpose machine. Not only does it remove for ever and completely all unwanted material, it actually pays for it in the form of energy; a powerhouse whose fuel is anything at all!

The picture of a black hole as a kind of energy machine is reminiscent of the situation confronting nineteenth-century engineers and physicists faced with the task of understanding the general principles governing the efficiency and work output of ordinary, down-to-Earth machines. A careful study of heat engines – devices which are able to convert heat energy into work or vice versa – led to the branch of science we now call *thermodynamics.* This subject is central to the proper understanding of the nature of time, and was discussed more fully in chapter 3. For the purposes of the present discussion, suffice it to say that a very general and fundamental principle was discovered, called the second law of thermodynamics. This law requires that the total entropy of a physical system shall not decrease. When applied to heat engines it determines the most efficient energy output to be those processes in which the entropy remains constant (called reversible processes). In the real world, processes always tend to increase the total entropy somewhat. This is an example of the time asymmetry discussed in the previous chapter. The entropy keeps on going up.

Black holes also possess a characteristic time asymmetry, as a result of the peculiar properties of the event horizon. Recall that this is a surface which only permits energy to cross *into* the black hole, never out again. Put simply, black holes apparently only get bigger, never smaller, as things drop into them. A quantitative measure of the size of a black hole is the surface area of the event horizon. Stephen Hawking has proved a remarkable theorem which says that the area of the event horizon should never go down, whatever processes the black hole indulges in. This theorem is a direct analogue of the second law of thermodynamics, with the event horizon area playing the role of entropy. It too may be used to place limits on the efficiency of black hole processes. One example of this concerns the process in which two identical Schwarzschild black holes are coalesced. An easy calculation shows that at maximum efficiency (no change of event horizon area) the total energy which can be extracted from the system is 29% of the original mass-energy.

Until a few years ago, these analogies between black holes and heat engines were really just a bit of fun for the theorists. Nobody actually expected to be able to use black holes as energy machines. Then came an astonishing discovery which not only confirms the relevance of the thermodynamic connection, but suggests new physical principles which might well extend far beyond the very narrow, academic confines of black hole theory, and throw new light on the nature of gravity itself.

At first, one of the flaws in the thermodynamic connection seemed to be that it was only an analogy. For example, the notion of *temperature*, so central to any discussion of heat, did not seem to apply very well to black holes precisely because they are supposed to be black. A completely black object has no temperature, i.e. its temperature is zero. The black hole ought to be completely cold, colder than anything else in the universe. The idea of a hot black hole did not seem to make much sense.

The first indication that black holes might not be completely black arose out of a study of an analogue of the Penrose process for energy extraction, but applied to light waves instead of particles. A similar energy enhancement occurs for light also, and is known as superradiance. In many ways superradiance is rather like the laser process. A laser (short for 'light amplification by the stimulated emission of radiation') is a device for causing atoms to emit light by stimulating them with more light of the same frequency. This process can only be properly understood through the laws of physics which apply to microscopic systems – called the *quantum theory*. A little bit about the quantum theory will be explained in the next section. The theory shows that atoms can also emit light radiation *spontaneously*, i.e. without first being stimulated. It is mainly by spontaneous emission that most objects radiate (for example, the sun).

It was suggested by the Soviet astrophysicist Ya. B. Zeld'ovich that if rotating black holes indulged in superradiant amplification of light energy they should also emit light spontaneously. This was the first suggestion that quantum theory should be applied to black holes. The challenge was taken up by the Canadian mathematician, William Unruh, who confirmed Zeld'ovich's conjecture and verified mathematically that a rotating black hole would indeed glow with a feeble light. Feeble is the word, for the energy radiated by this process from a solar mass black hole would be almost undetectable. Nevertheless, important issues of principle are raised, and it is essential that the mechanism which causes the radiation production is properly understood. The theory of quantum processes in curved space–time, which Unruh had to use, is still in a very tentative condition. However, the underlying physics may be sketched out for the curious reader.

The appearance of light radiation spontaneously from a rotating black hole means that some of the rotational energy has been converted into electromagnetic energy. The way in which this is achieved is, very roughly, visualised by imagining the space vortex around the spinning object. The violent dragging effect shakes up the electromagnetic field and generates ripples of energy – electromagnetic waves. It also shakes off neutrinos and gravitational waves (see page 128). The radiation being described here is not coming out of any kind of matter – the usual source of light energy. The region around the black hole is quite empty of matter. Instead, this radiation is coming *directly out of the empty* space itself! Thus, in a very real sense, the space whirlpool round a rotating black hole is not quite totally invisible. It glows with a feeble quantum light.

Significant though this result may be, it does not fully provide the missing temperature needed to clinch the thermodynamic connection. A Schwarzschild black hole does not rotate, so would not emit the Zeld'ovich–Unruh radiation. Nor does this radiation have a characteristic temperature, because its frequency spectrum does not correspond to that of a body in thermal equilibrium. The missing link was supplied by Stephen Hawking, who treated mathematically what appears at first to be a hopelessly complicated problem. He applied the quantum theory not to the end product black hole, but to the gravitational collapse process itself. During the implosion of a star, the same sort of disturbance of the electromagnetic field occurs, with ripples of light energy flowing out of the collapsing object. Once again, the energy is not produced directly by the matter of the star, but by the curved space–time. What is so remarkable about Hawking's result is that, as the star settles down (extremely rapidly remember) to form a black hole, so too the emitted radiation settles down to a steady flux of energy, which is quite *independent* of the details of the collapse process. Instead of getting a very complicated answer, Hawking obtained the simplest and most elegant possible result. The radiation from a Schwarzschild black hole has precisely the right frequency spectrum to correspond to a body in thermal equilibrium, at a temperature which depends only on its mass. Here was the missing link in the thermodynamic picture.

One consequence of the new result is that the laws governing black-hole processes simply become the usual laws of thermodynamics. The event horizon area *is* the entropy of the black hole. Consequently, this area can now decrease after all, without violating the second law, provided the entropy of the black hole environment increases by at least as much. This conclusion has a profound consequence. A black hole can grow *smaller*. Indeed, it *will* grow smaller if it is sufficiently small in the first place. The reason for this is the curious fact that Schwarzschild black holes get *hotter* as they radiate energy. The temperature which Hawking calculated depends *inversely* on the mass of the object. As the radiation flows away, the mass falls and the temperature rises, further encouraging the emission of radiation energy. The whole system is explosively unstable.

This means that a black hole will, if left long enough, evaporate completely away leaving, as it were, only the smile! The star which went to make up the black hole in the first place has apparently disappeared out of the universe in a shower of radiation.

For a solar mass black hole the temperature is a mere 10^{-6} degrees absolute. Such objects would grow rather than evaporate, as radiation from their hotter surroundings flows in at a greater rate than the Hawking radiation flows out. However, the microscopic black holes mentioned on page 110 would tend to evaporate over a time scale comparable with the age of the universe. Some of these mini black holes may be in their explosive death throws even now.

In this strange and fascinating subject it is as well to remember that we are here probing the very limits of current mathematical theory. The strong thermodynamic links discovered by Hawking and others do suggest that we are beginning to uncover important new principles about gravity and quantum theory. The new quantum results are very appealing and compelling. But we are very far indeed from any kind of direct observational confirmation of the sort of processes that have been described above.

Before we leave the subject of black holes something should be said about the possibility of our being able to observe one. If a very massive one exists at the centre of the galaxy it would probably eat up stars at a steady rate. The second law of black hole thermodynamics requires that the black hole can only grow in size, in the sense that matter falling inwards must cause an increase in the total horizon area. Nevertheless, calculations show that a fraction of the mass of the infalling matter could be radiated away in the form of gravitational waves. A gravitational wave is the analogue for the gravitational field of the electromagnetic wave for the electromagnetic field. Both travel at light speed, but whereas electromagnetic waves are produced by the disturbing effect on the field of accelerating charged particles, gravitational waves are produced by bodies with mass (gravitational charge).

In recent years a great deal of effort has been expended in constructing gravity telescopes to 'look' at sources of gravitational waves. The waves are detected in the simplest manner imaginable. The ripples of the space geometry caused by the fleeting passage of the wave are used to set into resonant vibration a metal cylinder. Cylinders in current use are a few metres in size and extremely delicately suspended to reduce terrestrial disturbances such as seismic waves. In spite of this, theoretical calculations show that even black-hole events are unlikely to produce sufficiently violent waves to enable current detectors to vibrate much above noise level, so that they are usually operated in tandem, and coincident movements noted.

The pioneer of gravity wave detectors is the American astrophysicist Joseph Weber. Great excitement was generated by his claim a few years ago to have detected strong sources of waves in the centre of the galaxy, which were widely believed to be some form of black-hole events. Since Weber's claim was made, other detectors in Europe and America have failed to confirm his results and it appears likely that we shall have to await the next generation of more sensitive detectors before the controversy is resolved.

There seems to be much more likelihood of the detection of a nearby black hole with stellar mass. Naturally, being black, such an object cannot be seen directly. Nevertheless, it still gravitates and there is a good chance that one might be found together with an ordinary star in a double star system. Binary stars, with two stars orbiting closely around a mutual centre of gravity, are fairly common in our galaxy. If one was a black

Fig. 4.17. Gravity telescopes. Gravitational waves could be emanating from black hole collapse at the centre of the galaxy. These 'ripples of geometry' can be made to set metal bars ringing. Several such devices are currently searching for these elusive waves, usually operating in pairs to distinguish local effects, such as earthquakes.

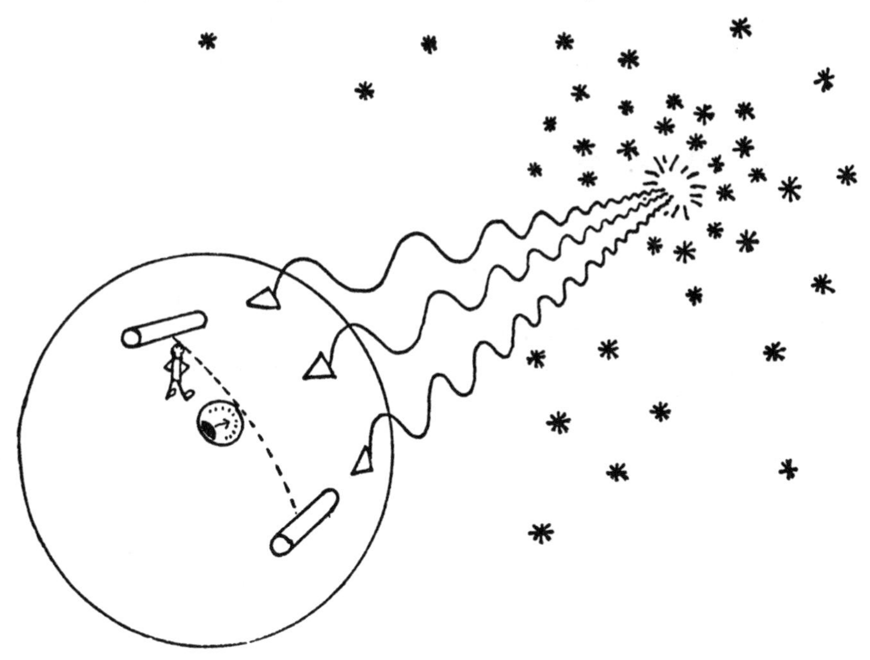

hole, it would tend to accrete material from its companion. This gas would probably form into a disk around the black hole, with the swirling matter slowly sinking down into oblivion. As it sinks, the gas would become very hot – so hot that instead of shining with visible light, it would emit X-rays.

Thanks to space technology, it is now possible to fly X-ray telescopes above the atmosphere in satellites. Recent flights have produced a number of possible candidates for black holes in binary systems, the most likely being in the constellation of Cygnus. In order to be definite about the black hole interpretation it is necessary to rule out the possibility of the radiating object being a white dwarf or neutron star. The object itself cannot be seen but its presence is inferred from the motion of the companion. The only criteria available for ruling out the more conventional contenders are based on theoretical models which predict that neither can apparently have a mass much greater than about a solar mass. The mass of the object can be estimated from the motion of the binary system provided the mass of the ordinary companion star is known. This can be inferred from its luminosity and the quality of its light.

However, a number of complicating factors muddle the estimation and as yet it can only be said that the presence of a black hole in the Cygnus object seems rather likely.

4.5 The fleeting worlds of the quantum

It is a remarkable historical fact that the early years of this century witnessed not just one, but two, major revolutions in physics, and indeed, in human thought. The theories of relativity modified at a fundamental level our conception of the large-scale properties of space, time and matter. Alongside the development of relativity, there grew out of the work of theorists like Max Planck (German, 1858–1947), Niels Bohr (Danish, 1885–1962), Erwin Schrödinger (Austrian, 1887–1961), Paul Dirac (British, 1902–), Max Born (German, 1882–1970), Werner Heisenberg (German, 1901–1976) and many others a strange new picture of the *small-scale* features of the world, a picture that in many ways is a more profound and more alien reconstruction than relativity. This revolutionary new view of the microcosmos is contained in that now vast branch of modern physics known as *quantum mechanics.*

Unfortunately, the theory of quantum mechanics is much more difficult and abstract than the theory of relativity, which prevents a lot of people without the necessary mathematical background from understanding its beauty and subtleties. It is only possible to give a rather vague outline of the essential features within the scope of this book. (The author fully appreciates that in attempting to popularise this subject certain phraseology and analogy might not accurately reflect every quantum physicist's conception of the subject. The reader should be cautious about drawing too many conclusions from a description which is necessarily heuristic.) Part of the difficulty of understanding quantum theory is the absence of accurate physical intuition based on familiar objects and concepts from the everyday world. In the theory of relativity, the notion of clocks and measuring apparatus, geometrical constructions and so forth are rooted strongly in daily experience. In contrast, the internal structure of the atom bears no real resemblance to the world of experience. Phrases like 'particles revolving round a nucleus' are often used in descriptions of atomic structure, but convey the impression that the only difference between the behaviour of atoms and macroscopic mechanical systems such as snooker balls is one of *scale*; this is quite false.

It was realised by the early part of this century that Newton's laws of mechanics (or, for that matter, Einstein's special relativity) could not successfully account for the behaviour of microscopic mechanical systems such as atoms. The simple mathematical laws which apparently describe so well the motion of snooker balls do not apply in this miniature domain. Nevertheless, atomic structure was known from experiment to possess symmetries and regularities of its own distinctive character, and a search began for a new type of mathematical description which could account for these experimental facts. By about the mid 1920s this mathematical framework was complete, and elevated to the status of a full new theory of matter, called quantum mechanics.

The success of the new theory was quite extraordinary. At a stroke it provided an accurate quantitative description of the following: atomic structure and scattering processes; the formation of molecules and chemical bonding between atoms; radioactivity and the internal behaviour of the atomic nucleus; the interaction of electromagnetic waves with matter (such as in the photoelectric effect); many of the properties of solids and many, many other laboratory phenomena. By the 1930s the quantum theory had been unified with special relativity by Paul Dirac and a whole new chapter of physics was opened.

The first pages of this new chapter are a classic example of the scientific method at work. From his mathematical theory of relativistic microscopic particles, Dirac was able to predict many new things. The model of the atom (well established by the mid 1930s) required the existence of three 'fundamental' constituents, or elementary particles, the proton, the electron and the electrically neutral neutron. Dirac was able to show mathematically from his theory that, for example, the electron is *spinning* in a fashion which is quite impossible in an ordinary macroscopic body. The effects of this spin are in fact a well-known feature of atomic spectra. More dramatically, Dirac also found that his relativistic matter equation, as well as possessing solutions which correctly describe the motion of ordinary electrons, protons and neutrons, in addition possesses 'mirror' solutions which apparently describe whole new types of particles. When Dirac first produced his solutions, no other particles were known. However, using a remarkable physical argument Dirac proposed a mechanism whereby the 'mirror' particles could be created, and in 1936 the first of these particles was found. It was the mirror of the electron – a particle with the same mass, but opposite electric charge. The new particle was christened the positron, and represented a spectacular confirmation of Dirac's relativistic quantum mechanics.

The discovery of the positron was only the beginning. In 1935 the Japanese physicist Hideki Yukawa predicted, on the basis of a theory of nuclear structure, the existence of another type of particle with a mass intermediate between the electron and proton, called the meson. A search did reveal such a particle, known as the muon, found in 1937. Only after the Second World War was Yukawa's meson, now called the pion, discovered. In addition to the mesons, the features observed in the process of β-radioactivity suggested another curious type of particle called the neutrino. It has no mass or charge, but spins in the peculiar fashion described by Dirac's theory. The interaction of the neutrino with other particles is extremely low – most of them would pass straight through the Earth without stopping – making it the most elusive particle known.

If the number of particle types had remained limited to the proton, neutron, electron, positron, the mesons, the neutrinos (two types are known) and the photon (quantum 'particle' associated with electromagnetic waves), then some hope would have remained of a description of truly elementary particles, out of which all matter is constructed. However, since the war, hundreds of new types of subatomic particle have been discovered, many of which only live for a fleeting moment (10^{-24} second, for example!). Although some unexpected symmetries have brought a little order to the chaos of these new entities, there is as yet no firm evidence of whether all these particles are made out of common elementary constituents, or whether their number and variety is unlimited. The search for, and characterisation of, these new particles is now a major industry involving millions of pounds of equipment. In spite of the obvious misnomer, this field of experimental and theoretical physics is known as elementary particle physics. The foundation stone of this discipline is based on special relativistic quantum theory.

To understand how the microscopic particles of this bewildering array are produced, recall that Einstein's special relativity formula $E = mc^2$ provides for the interconversion of mass and energy. In chapter 2 a description was given of how an accelerated particle becomes more massive as a result of this conversion. Another procedure for changing energy into mass is by the *creation* of new particles of matter out of energy. The energy can be supplied in several ways. One common technique is to simply smash existing particles together as hard as possible. Modern accelerating machines such as the one at the Centre Européen de la Recherche Nucléaire in Geneva are capable of producing whole showers of newly created particles of all types in this way.

By conducting experiments of this sort, elementary particle

physicists have discovered that all the different particles obey certain well-defined rules when they change into one another. These rules are characterised by assigning 'labels' to the particles; for example, electric charge and spin. The labels do more than just distinguish between the different particles. They are often *conserved* when the particles change in number or type. For example, a label called baryon number, which is carried by a neutron, will be passed on to the proton when the neutron decays into a proton, electron and neutrino. In contrast, a meson does not have baryon number, so that the decay of neutrons or protons into mesons is not allowed – and not observed. Also when a particle is created out of energy, the conservation of its labels usually requires the simultaneous creation of another particle which carries equal and opposite labels. Particle production therefore takes place in *pairs*. The creation of a positron, for instance, can only take place if an accompanying electron (with opposite charge and spin) is also created.

This requirement of label symmetry during creation (or destruction) implies that *every* particle possesses a Dirac-type 'mirror' image, with equal and opposite labels; not only the electron, but the proton, neutron, mesons, etc. should have their corresponding *anti-particles*. This is indeed so. The proton, for example, can be created with the accompaniment of an anti-proton, which has negative charge and baryon number. An anti-proton and a positron may together form an atom of anti-hydrogen. Matter in this mirror form is referred to as *anti-matter*. When matter and anti-matter meet, they immediately annihilate and convert back to energy, say in the form of gamma-ray photons. Consequently, particles of anti-matter do not last long in the matter-infested terrestrial environment, or indeed in our galaxy, which seems to be composed almost completely of matter. It is not known whether the entire universe is made of matter, or whether some anti-matter galaxies exist.

Having described very briefly some of the experimental consequences which have arisen out of the theory of quantum mechanics, some discussion about the extraordinary nature of the theory itself is in order.

The development of classical mechanics by Newton was based on the fundamental property of *predictability*. Newton constructed a set of mathematical equations to describe the time evolution of mechanical systems. In principle, provided sufficient information about the state of a physical system at a given time is supplied, its entire past and future history to arbitrary accuracy can be computed. A good example of this is the prediction of eclipses of the sun. The Earth–sun–moon system may be considered as a fairly straightforward, interacting three-body problem, and treated to a good approximation using Newtonian mechanics, because all the gravitational fields and velocities are small. Knowing the present state of the solar system then enables the dates of all past and future eclipse events to be calculated.

Thus, according to Newtonian mechanics, the world is like a machine, and the unpredictable only happens because we have insufficient information available to anticipate all events in detail.

About the turn of the century, Newtonian mechanics began to display shortcomings, because it could not describe properly certain vital properties of atoms and their interaction with electromagnetic radiation. There grew out of these shortcomings the entirely new conceptual structure of quantum mechanics. Quantum mechanics starts by denying the possibility of complete predictability in the world, irrespective of the amount of information available. Instead of regarding the present state of the universe as necessarily evolving into a completely determined future state (and from a completely determined past state), it conceives of many possible future and past states. The future state of a physical system is to be considered as a *superposition* of all possible outcomes. Rather than one future world there are stupendously many, and each of them may or may not occur with a definite, well-defined *probability*.

Prediction in physics, like economics, has thus become a matter of statistics. Whether we imagine that all possible worlds co-exist in parallel, or only one is selected at each moment in some random fashion, raises deep philosophical issues. In either case the physicist may calculate the *expectation* of any particular form of a physical system at any moment, using the laws of probability. For example, it is straightforward to calculate the relative probabilities that a nucleus of uranium has or has not decayed by the emission of an α particle. After, say, 1000 years, there are two possible types of world; one in which the uranium atom is intact, and one in which it has decayed. The theory of quantum mechanics provides a mathematical prescription for calculating the respective probabilities of each of these alternatives.

If we consider space–time on a small enough scale, it is possible for many different types of world to appear and disappear again, rather like ghost images. These brief manifestations are sometimes referred to as *virtual worlds* because they are only fleetingly glimpsed. Even if we start with a total vacuum – just empty space – over sufficiently short periods of time particles of all types come into existence and fade away again. The duration of these phantom virtual particles is incredibly short. A virtual proton, for example, can only live for about 10^{-24} seconds before vanishing again. Nevertheless, we can no longer think of a vacuum as 'empty'. Instead it is filled to capacity with thousands of different types of particles, forming, interacting and disappearing, in an incessant sea of activity. This is the quantum picture of space–time, one of violent fluctuations and interactions.

Nor is this quantum picture just an intellectual model. Very real physical effects occur as a consequence of this fluctuating vacuum. For example, the presence of electrically conducting materials will modify the fluctuations of the photons ('particles' of electromagnetic radiation) in a way that leads to measurable forces on the conductors.

This fascinating new quantum conception of space–time is a result of the fusion of quantum mechanics and special relativity. The unification of quantum theory and *general* relativity might be expected to produce equally drastic modifications in the space–time picture.

In recent years much effort has been devoted by theorists to understanding the nature of elementary particles in the curved space–time of general relativity. The little progress that has been made has only emphasised how strongly the existing notion of these particles is rooted in the flat space–time of special relativity. Nevertheless, recent work in the Soviet Union and the United States strongly suggests that the creation of elementary particles in strong gravitational fields might be important for the large-scale properties of the universe. Moreover, in the context of black holes, there is the prediction that such particle creation would cause their complete evaporation, as mentioned in the previous section.

At a more fundamental level, it is possible to apply the theory of quantum mechanics to the gravitational field itself, that is, to quantise space–time. This has been a very active and exciting area of research for many years now, and the problems both of mathematical technique and of principle are complex and deep-rooted. Gravitation manifests itself as space–time geometry, so that the quantum theory of gravity provides for a superposition

Fig. 4.18. Space–time breaking up. John Wheeler has suggested that at a mini-microscopic level, quantum effects are so violent that they start to tear space–time apart completely, creating a sponge-like structure of wormholes and bridges. This violent activity goes quite unnoticed by a subatomic particle, which can be pictured as larger than a wormhole in the same proportion as the sun is larger than an atom.

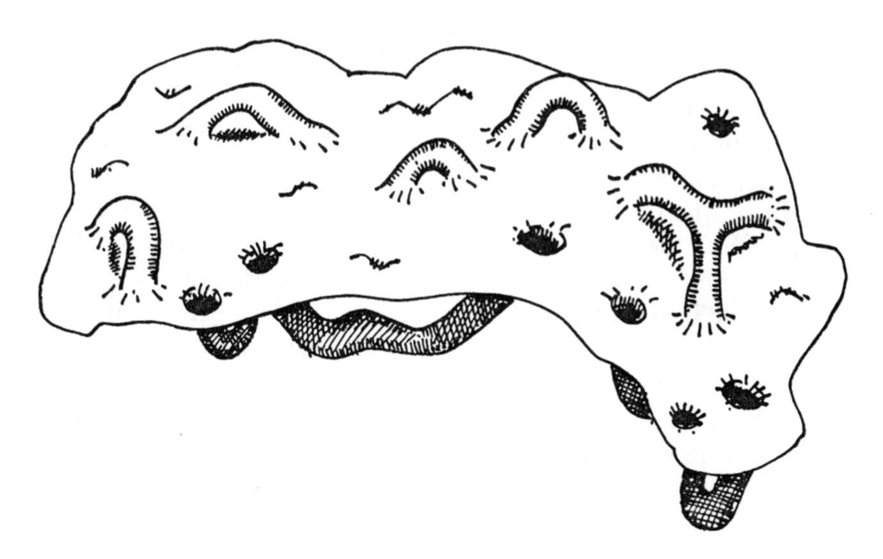

of worlds with different geometries. The observed geometry is to be calculated probablistically as usual. Once again, the vacuum is presumably full of fluctuations, but this time it is the geometrical structure which is fluctuating. On a mini-microscopic scale, virtual worlds with fantastically distorted and twisted geometries appear and reappear, form and decay, in ceaseless activity. According to some popular accounts, on the unimaginably small scale of 10^{-33} centimetre (twenty powers of ten smaller than an atomic nucleus) the fluctuations of space–time are expected to be so great that *topological* changes can occur. Virtual worlds with 'wormholes' and 'bridges' in space can then appear and collapse, giving space–time on this scale the properties of a sort of foam. In this strange region of foaming, virtual worlds all intuition about gravitational collapse and singularities breaks down. The quantum picture could lead to anything. In the absence of a proper theory there is no unanimous agreement about how seriously to take this sort of picture. All that can be said at this time is that the concept of a continuous space–time probably cannot be extended down to these extremely small regions.

A masterly proponent of this quantum 'geometrodynamics' is the American physicist John A. Wheeler. He has pointed out that the energy of all these fluctuations is so great that the presence of a real particle in this space is no more significant than a cloud is to the dynamics of the air. For although we notice the cloud – as we also notice the particle – it is but a minute disturbance on the underlying dynamical activity. With this view of space–time and matter we seem a world away from Leibniz and Mach, who tried to build space and time out of matter. Modern quantum theory gives space–time itself the central role, and regards matter as a mere disturbance on the underlying structure. It would be foolish to suppose that the story ends here. Doubtless a future theory will build together space–time and quantum theory in a much more basic way, and an entirely new concept of space–time will emerge. For the moment the reader must be content to regard relativity and quantum theory as a glimpse of the true reality from two of its many fascinating sides.

4.6 The current status of the general theory

In view of the fact that most of this chapter is based on the general theory of relativity, it is appropriate to finish with a few words about its status as a scientific theory. It has long been generally accepted by physicists as the best available description of space–time and gravitation. In large part, this acceptance stems from the truly beautiful subtlety and elegance of the theory itself as a description of nature. Unfortunately, this elegance is not extended to most applications of the theory, where technical problems of a mathematical nature render all but the simplest systems hopelessly insoluble. Part of the difficulty arises because gravitation is a form of energy, and is therefore its own source; put in a nutshell, gravity itself gravitates. Translated into mathematical language this manifests itself as a non-linearity of the equations of the theory. What this means is that the sum of the gravitational effects is not the same as the effect of the sum; one cannot just add together known solutions from simple systems to obtain the net solution of a complex system. As a result of these technical problems the true content of the theory is still being discovered, even after 60 years.

The experimental verification of the theory is extremely poor, nowhere near as good as the special theory of relativity. Besides the effect on Mercury's orbit and the red shift of light, experiments have been performed on the bending of light rays by the curvature around the sun. Agreement is adequate, but there are complicated additional factors in the experiment. There are a large number of alternative theories of gravitation, some are metric theories like Einstein's, based on a geometrical interpretation of gravitation, others are not. It is generally agreed that almost all of these alternative theories are excluded by the, albeit meagre, experimental data, whilst none has the aesthetic appeal of Einstein's theory.

In the quantum domain, there are some very remarkable arguments in favour of general relativity as the correct non-quantum limit. A consideration of just the elementary principles of the quantum theory of fields leads almost exclusively to the equivalence principle, geometrical transformation properties and non-linearity of gravitation. Nevertheless, the continuing failure to provide a truly successful marriage of quantum theory and general relativity has led to an accumulation of opinion that, in the quantum domain, Einstein's theory is wrong. Whether a modification of the existing theory can be happily patched together with quantum theory, or whether some radically new pre-quantum, pre-space–time structure will emerge is not clear at the time of writing.

5 Modern cosmology

5.1 The architecture of the universe

No analysis of space and time is complete without discussion of its totality. The totality of space is the universe, and the totality of time is the history of the universe. What can be said about the structure, life, birth and death of the cosmos?

It sometimes comes as a surprise that science can make any contribution at all to topics such as the creation and end of all things. Such matters are often considered to belong to the realm of religion or philosophy, and indeed, that is where they did belong throughout most of human history. Scientists are often asked what they *believe* about these issues; such a question fails to appreciate the considerable advance made in recent years in understanding the nature and evolution of the universe as a whole. It is no longer necessary for scientists to have *beliefs* about when the creation of the universe occurred or what form it took (though they do have beliefs), it is now a question of using scientific instruments to *see* what the universe is like and how it has evolved. Such great philosphical issues are not debated as acts of faith, but as questions of *evidence* and *theory*, in the same way as other scientific disciplines. It is true that much of the present understanding of cosmological matters is rudimentary and tentative; certainly great upheavals in the currently accepted picture of the cosmos are very likely in the future. Nevertheless, it is important to appreciate that we are dealing here with science, and scientific values, so that while personal religious or philosophical preferences may make a great contribution to a particular individual's conception of the universe, the topics to be discussed in this book deal solely with concrete observational data and the controversies which rage around their theoretical interpretation.

It is as well to begin with a general description of the structure of the universe as it is understood today. Perhaps the most conspicuous feature is the emptiness; nearly all the universe is empty space. Of course, this is only true in a sense. Quite apart from the quantum considerations mentioned at the end of chapter 4, there are always present certain amounts of radiation and residual atoms. But for intuitional purposes the material contents of the cosmos should be imagined as separated by vast distances of space.

Most of the luminous matter is in the form of *stars*. Each star is similar to our sun, though their sizes, colours, compositions and evolution vary appreciably. Stars are grouped together, along with some dust and gas (and other objects besides), into gigantic clusters called *galaxies*. A typical galaxy such as our own contains about 100 billion stars (about the same as the number of cells in a human brain) and measures about 50 000 light years across.

In human terms a galaxy is unimaginably huge. Yet on a cosmological scale it is insignificantly small. Galaxies are spread throughout the universe in a more or less random fashion, except that usually they are clustered into small groups. These groups of galaxies are the 'atoms' of cosmology. Statements about the behaviour of the universe refer only to scales of this order and above. The activities of individual galaxies, whilst they may be of great interest, rarely have a bearing on the subject of cosmology.

The reader may find it helpful to relate this hierarchy of structure to what he can actually see when looking up at the night sky. Apart from the sun and moon, the brightest permanent objects in the sky are the nearby planets. They belong to a group of nine worlds which includes the Earth, and which orbits around the sun (which is 700 000 kilometres in radius – 100 times greater than the Earth) at distances of up to a few billion kilometres. This group of planets together with the sun is called the *solar system*. Although similar in appearance to stars, these other planets are really very small and dim in comparison. They are only visible because they reflect the light from the sun, and being so close to the Earth this reflected sunlight seems quite brilliant. However, even our largest telescopes if transported to the nearest star would not be able to discern these tiny companions of the sun.

In contrast, the stars themselves are large incandescent suns, so far away in comparison to the planets that even though they shine millions of times more brightly, they appear to us to be relatively dim. It is expected that most other stars have their attendant planetary systems, similar to the solar system. Something about the structure and evolution of stars was mentioned in chapter 4.

The stars which are visible individually in the night sky are just the nearby members of our galaxy. Most of the galaxy is only visible as a blurred band of light passing right across the sky, known as the Milky Way. Moderate telescopes will resolve the Milky Way into myriads of separate stars. The centre of the galaxy lies in the direction of the constellation of Saggitarius (though well beyond).

Other galaxies are not readily visible to the naked eye, though a handful may be seen with powerful binoculars. The Andromeda galaxy, which is large and nearby (a mere one-and-a-half million light years away), is just visible to a person with good eyesight as a fuzzy patch in the constellation of Andromeda. Modern telescopes are capable of detecting hundreds of millions of other galaxies. Their structures vary greatly, but one distinctive and beautiful shape is that of a flattened disk with a bulbous centre and emanating spiral arms, rather like a catherine wheel. Both the Andromeda galaxy and our own are of this spiral type. The position of the solar system in our galaxy is about two-thirds of the way from the centre, in one of the spiral arms.

It should always be remembered that when we look out into the universe we do not see the galaxies as they are now, but as they appeared in the distant past. This is because the light from them has to travel millions and millions of kilometres to reach us, and this journey may take millions of years. Light from the *nearby* Andromeda galaxy takes one-and-a-half million years to reach us. Large telescopes can see much more distant galaxies that appear as they were *billions* of years ago!

Although only the galaxies are seen through large telescopes, there is undoubtedly *some* material in the dark spaces between them. Just how much, or what the nature of its physical condition might be, is still a matter of conjecture. In addition, various types of radiation and particles apparently permeate the whole universe. These include electromagnetic and gravitational radiation, neutrinos and cosmic rays (consisting of many different types of subatomic particles).

Having outlined the distribution of matter in the universe a few words should be said about what the matter is made of. All matter is composed of *atoms*. Ninety-six different kinds of atoms occur naturally on Earth; in addition, several new types have been made artificially. A substance composed entirely of a particular type of atom is called an *element*. Atoms of most elements can combine together with other atoms of the same or different elements to form *molecules*, the precise rules of combination being the subject of chemistry. All matter in its various forms, from diamonds to air, human beings to stars, are made up of different combinations of these same basic elements. The simplest element is hydrogen. Its atoms consist of just two particles, an electron and a proton. The next simplest is helium which has six particles; two protons and two neutrons stuck together at the centre forming a *nucleus*, and two electrons in orbit about the nucleus, bound by electric attraction. The most complex naturally-occurring element in abundance is uranium which has 96 protons and about 140 neutrons in the nucleus, with 96 electrons spinning around it.

The chief differences between atoms are due to the varying number of protons in the nucleus. All possible atoms from one proton to 96 protons are known on Earth, but whereas some, such as iron, are very abundant, others, like technetium, are exceedingly rare. Those elements with more than 96 protons which have been made artificially (such as neptunium and plutonium) are unstable (radioactive) and fall apart rather easily. That is why they are not found occurring naturally on Earth.

Spectroscopic studies of astronomical objects show the presence of these same elements. Indeed helium, as its name suggests, was known in the sun before it was discovered terrestrially. However, the relative abundances of the elements on the Earth are not at all typical of the wider universe. Estimates show that about 90% of all atoms in the universe are hydrogen. Helium accounts for most of the remainder, and all the heavier atoms, so common on Earth, form a very small fraction of the total. Clearly a strong selection effect was at work when the Earth was formed.

The rapid decline in the abundance of the elements with increasing weight strongly suggests that the universe began without complex atoms, and that a *synthesis* mechanism has been at work, building up the complex elements out of the lighter, simpler ones, such as hydrogen. Just where the element-building factory is located will be discussed in due course. It turns out to have the greatest relevance for the *time asymmetry* of the universe. The question of where the hydrogen came from in the first place is a searching one, which will be expanded upon in section 6.1.

Undoubtedly the most significant feature about the universe is the high degree of *uniformity*. This is true in two distinct senses. Firstly, the detailed structures of distant stars and galaxies, the laws of physics which they obey, and the naturally-occurring quantities (such as the charge on the electron), all seem with great accuracy, to be the *same* as those encountered in our particular neighbourhood of the universe, and indeed on Earth. A typical galaxy hundreds of millions of light years away looks much the same as our own. The spectra of its atoms, and hence the chemistry and atomic physics there, are identical to the terrestrial case. This fact alone gives great confidence in applying laws of physics discovered in the laboratory to the wider universe.

The second aspect of cosmic uniformity concerns the distribution of matter. From the description of the universe just given it is clear that this distribution is very clumpy. The matter is gathered up into stars, which are arranged in clusters up to the size of galaxies. These galaxies are clustered together into groups. A few cosmologists claim that this clustering continues indefinitely in a hierarchy, the progressive clusters being separated by ever greater chasms of empty space. Generally, however, it is assumed, on the basis of some reasonably good observational evidence, that the clumpiness stops with the galactic clusters, and thereafter, the large-scale distribution of matter is very even throughout the universe. This distribution is both homogeneous (the same in every region) and isotropic (the same in every direction). Not only is it aesthetically satisfying that the large-scale structure of our world is so very simple, but also it is extremely convenient theoretically, for it enables mathematical models of the universe to be constructed with a minimum of technical complexity. In addition, this structure is in accord with the modern post-Copernican philosophy that the Earth does not possess any privileged status in the universe. Before Copernicus, European thought placed the Earth at the centre of the universe, with all other heavenly bodies revolving about it. Copernicus' discovery that the Earth moves around the sun shattered that illusion for all time, and humanity has never quite recovered from the shock. These days it is considered mildly heretical to assume that our region of the universe is anything but a *typical* region. The physical conditions in our neighbourhood are considered no longer special, but representative of any average place in the universe. Our Earth, sun or galaxy may seem to be of extreme importance for human beings, but in the total scheme of things they are quite insignificant and unexceptional.

The assumption that the universe is uniform on the large scale is now accepted by most (though not all) cosmologists, and is known as the *cosmological principle.* Carried to its extreme, the post-Copernican philosophy could be invoked to conjecture that not only is our region of the universe a typical part of space, but also that our present epoch is a typical *time.* This would imply that, by and large, wherever and whenever the universe was inspected, it would look more or less the same.

A picture of the universe along these lines was widely accepted by astronomers a century ago. Constantly burning stars were envisaged as scattered uniformly throughout all of infinite space.

This awe-inspiring and somewhat sterile model suffered, however, from a number of serious shortcomings. One of these has generally become known as Olbers' paradox after the nineteenth-century German astronomer Heinrich Olbers (1758–1840). It centres around the inconsistency which arises between an infinite, unchanging universe and the darkness of the night sky. It may seem a trifle fatuous to enquire why the sky is dark at night, but in the context of this model universe there is a very real problem here. Modern physics best exposes this problem with thermodynamic language. A dark sky is a cold sky, so we conclude that the universe is on average very cold (it is in fact about three degrees above absolute zero). In contrast, the stars, like the sun, are very hot. Their surfaces have a temperature of thousands of degrees, while their interiors may reach hundreds of millions of degrees. Why then, put simply, have the stars not heated up the universe by now? How can an unchanging universe always be in thermodynamic disequilibrium? If radiation has been pouring out from the stars for all eternity, the universe ought to be a very hot place, and the night sky should be filled with radiation at a temperature of thousands of degrees. If it were, we should be instantly vaporised.

A century ago the science of thermodynamics was not well developed, and Olbers expressed his paradox in terms of the theory of optics. His reasoning is easy to follow. If the universe is infinite in size and unchanging in time, with a uniform density

Fig. 5.1. Olbers' paradox. If the stars continue outwards through space for ever, at a uniform density, every line of sight will eventually strike one. There can then be no dark parts in the sky.

Notice that although the more distant stars appear fainter, they also appear smaller in the same ratio. Each star therefore lights up the patch of sky that it does occupy with the same brightness. As there are more distant stars than nearby ones, the total fraction of sky occupied by all the stars at a particular distance does not, as it turns out, depend on this distance. Thus, if we can see far enough, the entire sky will be totally covered by stars. So why is the sky dark at night?

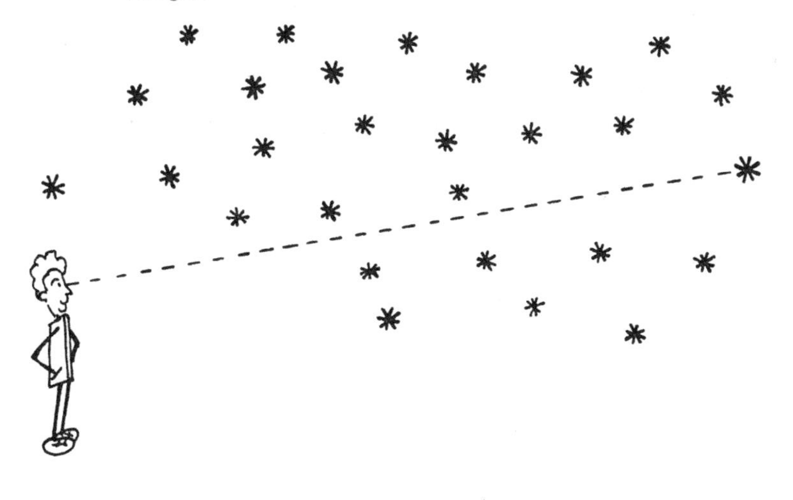

of constantly shining stars, then everywhere one looks in the sky a star should be seen. Every line of sight would, if continued sufficiently far, intersect a star. There could then be no dark parts in the night sky. Every direction would shine with the intensity of a star.

It is not hard to think up a number of *ad hoc* resolutions of Olbers' paradox. For example, if the stars did not continue to infinity after all, but were gathered together in a large blob in otherwise infinite, empty space, then all the excess radiation could flow away into the void beyond, though none would flow in. Naturally this model is not in the spirit of Copernicus, because the stars near the edge of the blob are in a special, rather than typical, position (they look out into empty space one way). It would have to be accepted as simply a coincidence that we were

Fig. 5.2. Cosmic catastrophe! A finite universe would have a middle and fall towards it, pulled inwards by the gravity of the stars. In 1692 Isaac Newton wrote that such a universe would 'fall down into the middle of the whole space, and there compose one great spherical mass. But if the matter was evenly disposed throughout an infinite space . . . some of it would convene into one mass and some into another . . . And thus might the sun and the fixed stars be formed.'

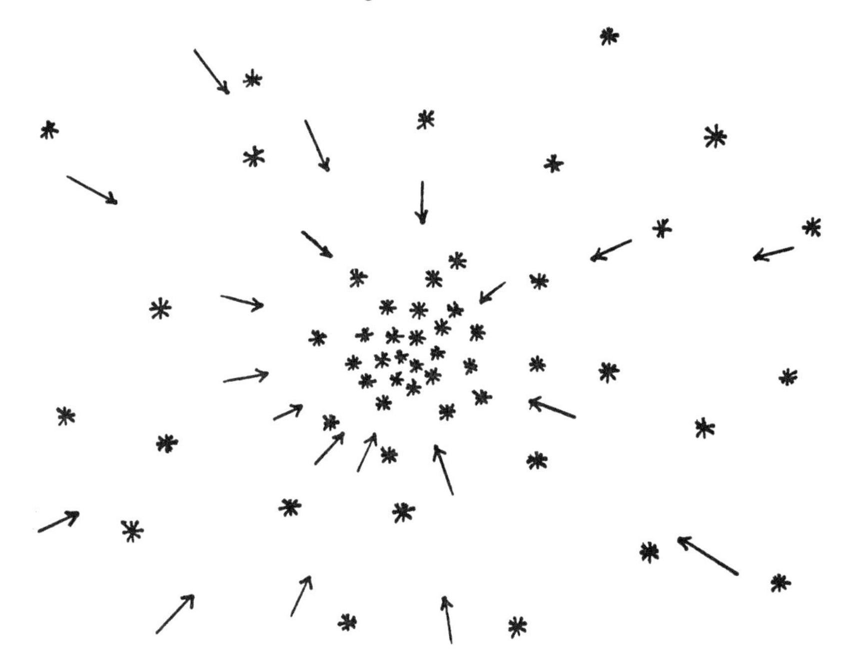

near the centre. A more serious objection had, however, already been pointed out by Newton. A force of gravitation attracts all the stars to each other. If the blob of stars has an edge, it also has a middle, so all the stars would fall down into it. Newton reasoned that as this had not happened the stars must extend to infinity.

A completely different resolution of Olbers' paradox follows from a remark made by Boltzmann, who suggested that the present thermodynamic disequilibrium of the universe was the result of a gigantic chance fluctuation (Poincaré cycle – see section 3.3), such as occurs every $10^{10^{80}}$ years or so! Left for this length of time, it would happen that now and then all the heat in the universe would flow spontaneously into the stars and raise their temperatures to millions of degrees. What we observe now is the fluctuation undoing itself and returning the stars to equilibrium with space. According to this proposal, the sky is dark at night simply because all the heat has happened to flow into the stars in orchestration. The reason why the human race is chosen to be a witness of this unimaginably rare event is precisely because living creatures, including cosmologists, require the disequilibrium (e.g. the presence of sunlight) thereby produced in order to survive.

In fact, Boltzmann's novel idea cannot be taken seriously. There is no reason why *all* regions of the universe need to fluctuate together to produce life on Earth. Had the disequilibrium arisen this way, it is overwhelmingly unlikely that when a telescope is turned on a distant region of the universe it would show brightly shining stars there as well. Such a universal fluctuation is vastly more unlikely than a local one.

The correct resolution of Olbers' paradox might well have been guessed by any one of the nineteenth-century astronomers. In the event, it had to await the arrival of the 100 inch telescope at Mount Wilson observatory in the USA, and an epic discovery as significant as that made by Copernicus.

5.2 The expanding universe

In the year 1929 the American astronomer Edwin Hubble (1889–1953) announced some results concerning measurements made on the light coming from distant galaxies. An examination of the frequency spectrum of this far-off starlight revealed that the spectral lines were systematically displaced towards the colour red (low frequency end of the optical spectrum). Hubble found that this so-called red shift increased in proportion to the *distance* of the galaxy. In chapter 2 it was explained how a frequency shift of light can occur as a consequence of the *recession* of the source of light, i.e. the Doppler effect. Clearly, distant galaxies are receding from us in a systematic pattern of motion; the farther away they are, the faster they recede. The conclusion is inescapable: the universe is in a state of *expansion*. This totally unexpected discovery changed the whole complexion of the subject of cosmology. An expanding universe is a *changing* universe, with a life history – perhaps even a birth and death. Paradoxes such as that of Olbers' are swept aside, for there is no reason to expect an evolving cosmos to be in thermodynamic equilibrium anyway.

In addition, the expansion raises an exciting new possibility, with a fundamental issue of principle associated. If the universe *moves*, does it obey some *law* of motion, like Newton's laws? Is it meaningful to treat the universe as a whole in the fashion of a dynamical object?

Most modern cosmology is built on the premise that the answer to these questions is in the affirmative. The global motion of the cosmos is assumed to obey the same laws that govern its individual component parts.

The next step is to decide what *force* regulates the cosmic motion. Only electromagnetic and gravitational forces are long-ranged enough to act across the enormous distances involved. In large objects, gravitation far outweighs electromagnetism in strength, even on the scale of the solar system. A theory of motion of the universe, therefore, is taken to be a theory of gravitation. By the time that Hubble made his discovery, the general theory of relativity was already established as the accepted description of motion under gravity. Accordingly, physicists started to study cosmic dynamics by constructing *relativistic* mathematical model universes.

Actually, the application of general relativity to cosmology had already been made by Einstein himself before Hubble's discovery. Curiously, Einstein was dismayed to find that his theory only gave rise to expanding or contracting universes. In keeping with the general belief at that time that the universe was unchanging, he was attempting to construct a *static* model universe, which would not fall together under its own gravity, or have to expand to escape it. He even went as far as modifying the general theory to meet this requirement, by adding an additional cosmic force of repulsion to balance the attracting gravity of the stars.

Einstein's model differed from earlier static models, based on Newton's theory of gravitation, in one novel and fascinating way. The Einstein model universe is *finite*, but still the same everywhere. That is, it is a universe limited in size but without an edge! Such a monstrosity is clearly impossible with Newton's model of space and time. However, the curved space of the general theory of relativity does admit this possibility. An

Fig. 5.3. On seeing the back of one's head . . . Einstein's model universe is finite in size, but has no boundary. Light may travel right around it in any direction, thereby returning to its starting point. This enables a person (with a sufficiently powerful telescope!) to see the back of his own head. The drawing shows a sphere. The spherical surface is a geometrical arrangement of two dimensions which reproduces this strange property of the Einstein universe.

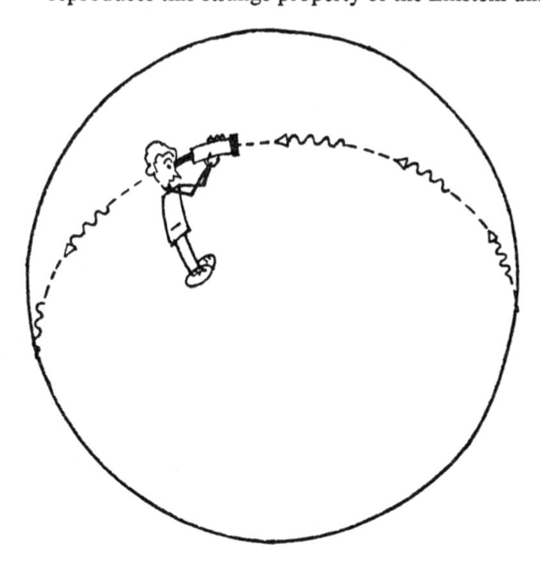

example of such was given in chapter 4, based on an analogy with the two-dimensional surface of the sphere. The spherical surface is finite, but has no edge or boundary anywhere – a finite, unbounded space. In Einstein's model, three-dimensional space also has a topology like that of the sphere, only in three dimensions rather than two, of course. Thus the Einstein universe has a finite volume of space, with the galaxies distributed evenly throughout it, in accordance with the cosmological principle, having an edge or boundary nowhere. Instead of continuing outwards for ever, space closes up on itself just as the surface of a sphere 'joins up round the other side'. What this means is that an inhabitant of such a universe could send a light signal out in any direction only to find that it would return again, from the opposite direction, having travelled all around the universe. One can envisage a space traveller on a journey in similar fashion – a sort of cosmic Magellan, circumnavigating the universe.

The idea of a closed, finite and unbounded universe was certainly a strange new idea of Einstein's. People often have a little difficulty in envisaging such an entity, a frequent question being what is 'outside' a finite universe. This question is as meaningless to three-dimensional humans as the question of what is 'outside' the surface of a sphere would be to a flat creature permanently restricted to live in the spherical surface. There can be no outside to the Einstein universe, because if there were an outside and an inside, there would have to be a boundary between them. No such boundary exists in this model. All points are equivalent to all others; none are near the 'centre' or 'edge'. There is no centre or edge.

The first person to use the general theory of relativity to construct a range of mathematical models of an *expanding* universe was the Russian meteorologist Alexander Friedmann (1888–1925) who published his work unobtrusively in 1922. These models remain the basic theoretical framework for the discussion of nearly all modern cosmology. The vital feature of Friedmann's models is their assumption of spatial uniformity. According to the cosmological principle the clusters of galaxies are assumed to be spread evenly throughout space. Friedmann assumed an exact uniform distribution of matter, and then

Pedestrian's guide to Friedmann's solutions

What goes up must come down! Not true, as even Newton knew, and our space rockets verify. Project a mass fast enough upwards and it will escape the Earth's gravity.

$$\text{Total energy} = \text{kinetic energy} + \text{potential energy}$$
$$= \text{constant.}$$

For spherical symmetry this well-known equation reads $\frac{1}{2}mv^2 - GM/R = E$ (constant total energy) where m and v are the projectile mass and velocity, R its distance from Earth and M the mass of the Earth. Note that v is the rate of increase of R, so this equation is easily solved for R as a function of time t. If E is positive, graph 1 of fig. 5.5 results; negative, graph 3; zero, graph 2.

These easy solutions are identical to the Friedmann models of the universe, calculated from Einstein's general theory of relativity – the most mathematically intractable theory of physics! In graph 1 the universe (projectile) escapes with energy to spare. Graph 2 is the limiting case of 'narrow' escape. In graph 3 the universe (projectile) falls down again.

---

solved Einstein's equations of general relativity for such a matter distribution, to find out how the geometry of space would change with time. Because of the uniformity, the only geometrical change which can occur is an overall change of *scale*, i.e. an expansion or contraction which is the *same everywhere*.

One popular way of depicting this expansion is by analogy with a rubber sheet. Such a sheet is shown in fig. 5.4 covered

Fig. 5.4. The expansion of space. The rubber sheet covered in dots is expanded uniformly. From any given dot, such as A, all other dots appear to recede – the distant dots faster than the nearby ones. There is no *centre* to the expansion, merely a universal change in the scale of all distances (e.g. R).

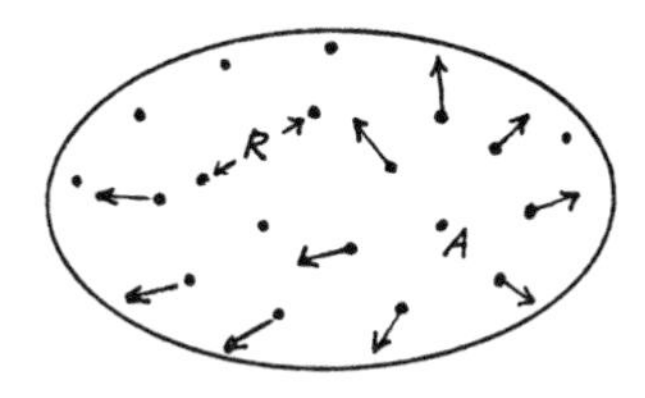

uniformly with black dots. These dots represent galaxies (or strictly speaking clusters of galaxies) and the rubber sheet represents space. The expansion of the universe can then be pictured as a stretching of the rubber sheet. To render the expansion uniform the stretching must be equal in all directions and at all points. As the sheet expands, so every dot moves away from every other dot. From the standpoint of any given dot, all the others seem to be receding from it, so that it appears to be at the *centre* of a general expansion. This is obviously not true, for all such dots witness the same phenomenon. There is no centre of expansion and no centre of the universe. Of course the rubber sheet as drawn has a centre, but this could be circumvented by making the sheet infinitely large, or by folding it over to form a sphere.

It should also be clear from this analogy that the expansion of the universe is one of *space itself*, and must *not* be pictured as the migration of the galaxies out into a pre-existing void. The

Fig. 5.5. The Friedmann cosmological models. As the universe expands, the size of a typical region of space grows. Above is a graph of this growth, showing the three possibilities discovered by Friedmann. All of them start out from 'nothing' at $R = 0$, models 1 and 2 expanding for ever, but 3 slowing to a halt, followed by collapse back to nothing again.

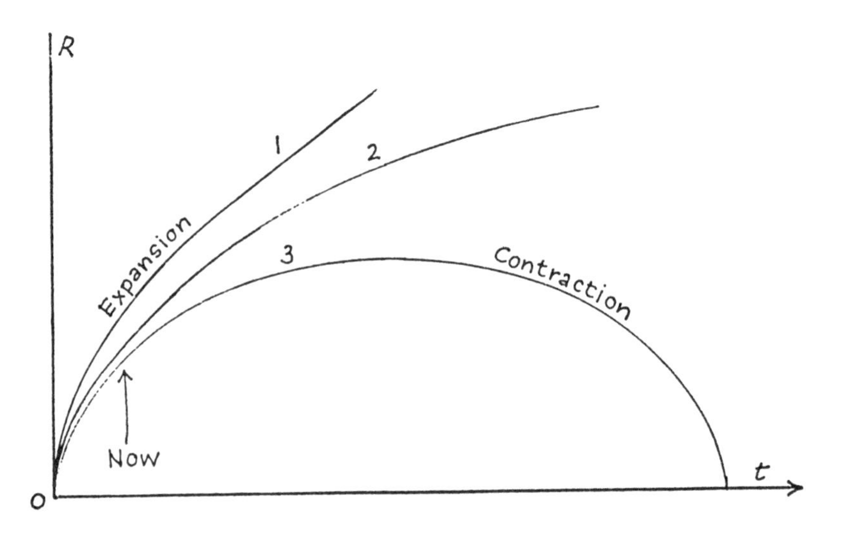

changing geometry can be characterised by specifying the distance between any two typical dots. Because the expansion is uniform, this distance will change its scale in the same way whichever two dots are chosen. It follows that the *rate* of separation of two dots is simply proportional to the distance between them, exactly as Hubble suggested for the galaxies. This fact alone gives some confidence that one of Friedmann's mathematical models is a good approximation to the large-scale structure of the real universe. But which model? Three possible models are represented in fig 5.5, which shows the scale factor, called R, plotted against time t.

Before these models are discussed in detail, a few words must be said about the nature of the time being used here. It should be clear from chapters 2 and 4 that the time used by any particular observer depends upon his relative motion and the gravitational field in which he is situated. How then, is it possible to construct a common time t to describe the behaviour of the whole universe which is itself in motion and changing its gravitational field? The answer hinges once again on the cosmological principle. Because the (large scale) universe looks the same from every galactic cluster, and changes by expansion at the same rate everywhere, the effect on clocks is the same at all places, provided of course that the clock is not in rapid motion relative to the local group of galaxies, when there would be a relativistic time dilation effect. The galaxies everywhere define a set of privileged reference frames – the frames in which the universe expands isotropically (the same rate in all directions) – and it is the clock rates in this privileged set which can meaningfully be compared. For example, the Earth is moving only slowly (compared with light) relative to the local group of galaxies, so that Earth time is an accurate way of dating the large-scale condition of the universe as seen by any distant observer travelling with his local galactic group. On the other hand, an observer in fast rocket motion past the Earth would have a different time scale of cosmic events. He would not belong to the privileged set of observers because his rapid motion would result in some galaxies in the direction of his motion appearing to approach rather than recede.

Table 5.1. *History of the universe: above, looking back in time from now; below, looking forwards from the big bang*

Feature	*Age in years*
Technological culture	100
Civilisation	10 000
Mankind	5 million
Mammals	200 million
Terrestrial life	3 billion
Earth	$4\frac{1}{2}$ billion
Universe	10–20 billion

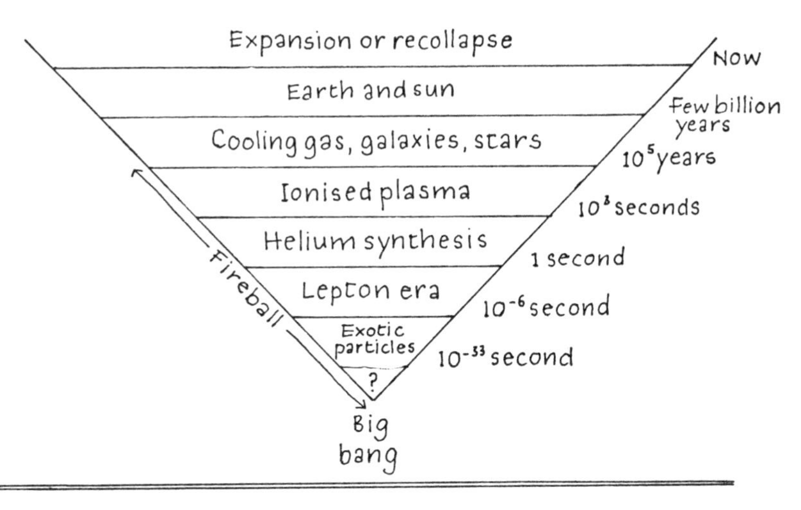

This universal clock time is called cosmic time, and because it happily coincides closely with Earth time it enables us to compare historical events on the Earth with various cosmic events. Such a comparison is given in the table above and might help the reader obtain some feeling for the *lengths* of time being discussed in this chapter.

To return to the Friedmann models, the three examples illustrated in fig. 5.5 are obtained by solving Einstein's field equations in which the *pressure* (which is a source of gravitation in general relativity) of the contents of the universe has been neglected. This is a good approximation at the present epoch, because the gravitating effect of the mass of all the matter in the galaxies is much greater than that of the small pressure in the universe (mainly due to radiation). A significant feature of these models, which is characteristic of a wide range of models in which a reasonable behaviour for the material contents is assumed, is that the expansion *rate* steadily decreases with time. All the curves shown in the figure gradually bend downwards. The expected position of the present time is indicated on the graph. It follows that at some finite moment in the past, the scale factor R must have been zero according to each of these models. This is an expression of the physical fact that the galaxies which are now seen to be moving apart must have been close together in the past. The precise condition of the universe near $R = 0$ requires some discussion which will be deferred until other features of the Friedmann models have been considered.

A glance at fig 5.5 also enables us to predict the future of the universe. Two quite distinct possibilities are indicated. In the models marked 1 and 2 the cosmological expansion continues for ever. An easy way of understanding the reason for this is as follows.

Fig. 5.6. Possible geometry of the expanding universe. Once more two-dimensional sheets mimic some of the features of three-dimensional space. In (*a*), the 'expanding balloon' corresponds to Friedmann's model 3 universe, which eventually recollapses. (*b*) is the 'flat' (Euclidean) model 2, in which the geometry is that which we learnt at school. In (*c*) the space curvature is 'outwards', in contrast to the inward curvature of (*a*). Both (*b*) and (*c*) are infinitely extended, but (*a*) is finite.

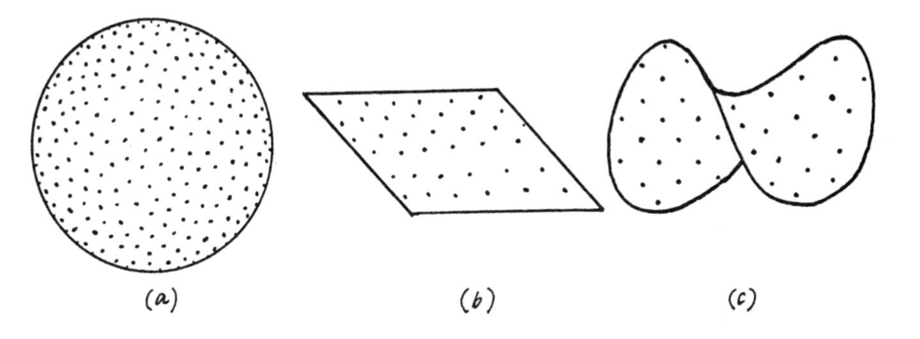

In model 1 the density of matter is so low that although the self-gravitation of the universe slows the expansion down somewhat, after a characteristic time the expansion is essentially free – the galaxies have 'escaped' each other's gravitation. In model 2 the density is sufficiently high to continue slowing the expansion for ever, but not great enough to ever stop it completely. In model 3, however, the density is high enough to actually reverse the expansion at a maximum value of R, and to drag all the galaxies back on themselves. This universe, like the Newtonian 'blob', then falls together to end in a condition similar to that at the beginning of the expansion.

The reason for the existence of three distinct models here reflects the fact that the space geometry of a uniform universe can be of *three* different forms. Roughly speaking, these forms of geometry are either with space curved inwards, like a sphere, or outwards like a saddle, or flat in the 'usual' way (see fig. 5.6). In the high density, recontracting model, space is curved inwards and, like the Einstein model, is finite in volume. In the Friedmann case, however, it is an expanding and contracting spherical space, rather like a balloon being inflated, then deflated.

The model denoted by the middle curve (2) in fig. 5.5 possesses a flat, Euclidean space, while the 'escaping' model 1 possesses outward curvature. Both of these models are infinite in volume. The idea of an infinite distribution of galaxies expanding is sometimes difficult for people to grasp. If the galaxies already fill all of space, what is there left for them to expand into? The reader should remember that the galaxies are not expanding *through* a fixed space, but are trapped *in* an expanding space. It is simply that the scale of all distances is everywhere increasing.

5.3 The creation of the universe?

If the universe is expanding, it must have been more compressed in the past. From the graphs shown in fig. 5.5 it appears inevitable that at some moment in the finite past, the scale factor of distances vanished at $R = 0$. This point represents the *beginning* of the expansion in the Friedmann models. The precise time when this occurred depends to some extent on which of the three models is chosen. A measurement of how fast the galaxies at a given distance are receding from us as this moment enables the expansion *rate* to be calculated, which determines our present position on the graphs shown in fig. 5.5. It is seen that the past behaviour of all three models is in fact very similar, and the moment when $R = 0$ turns out to lie between 10 and 20 billion years ago.

Recalling that R measures the distance scale between any two galaxies, the point $R = 0$ corresponds to a situation in which all galaxies meet together, and all lengths in the universe are collapsed to zero. Taken to the extreme this means that all the volume of space that we see, even through the largest telescopes, is shrunk, quite literally, into nothing.

It follows that all the matter in the observable universe, that now goes to make up the galaxies, with their millions of stars, the dust and gas, the intergalactic matter – was at that moment squeezed into a single mathematical point of infinite density! This is called a singularity in modern general relativity. The singularity was discussed in chapter 4 in connection with black holes. In fact, the condition of the Friedmann model universe at the beginning of the expansion is the same as that at the centre of a Schwarzschild black hole. How seriously should the singularity be taken? Recall from chapter 4 that the singularity is not really part of the theory. If the density of matter becomes infinite, Einstein's field equations can no longer meaningfully describe the situation. What the existence of a singularity in the Friedmann models implies is that general relativity, and possibly even the space–time description itself, must break down at a sufficiently early stage. Of course, it is known that the quantum theory of the gravitational field will become relevant when the length scale considered is small enough. In the case of the universe, this does not happen until the entire contents of the observable universe are squashed into a volume the size of a single atomic nucleus, which occurs at a mere 10^{-43} second after the initiation of the expansion. One thing is clear enough though; the space–time description cannot be continued through a singularity.

If space–time cannot exist at a singularity, then the point $R = 0$ in the Friedmann models describes a situation where space–time *came into existence.* The occurrence of the Friedmann singularity in this theory has therefore led to the widespread belief that the beginning of the expansion represents the *creation of the universe.* Certainly the singularity in general relativity is the nearest thing yet to the act of creation that science has discovered. If a singularity really did occur in the fashion described by the Friedmann models, with the matter density becoming infinite, then it is not possible to continue physics, or physical reasoning, through it to an earlier stage of the universe. That is to say, that nothing of physical relevance to the observed universe could occur before the beginning of the expansion, a condition which would seem to meet the requirements of an act of creation.

If the Friedmann models are taken literally, then not only space–time, but all the matter in the universe comes into existence at the singularity. Incidentally, this is the only place where the creation of pure matter is considered allowable by elementary particle physicists. It was mentioned in section 4.4 that particles carry various types of labels, which are normally preserved when these particles transmute into one another, or when they are created in pairs. This forbids the production of matter without a corresponding quantity of anti-matter. However, at a singularity all physical rules such as this break down, and matter may be created there unaccompanied by anti-matter.

An important feature of this scientific picture of the creation is that what is being created is both matter and the whole of space–time as well. This is in contrast to the biblical account of creation, where material things are created in a pre-existing void. Before the expansion there was not only no matter, but also no space or time either. The singularity must be regarded as a temporal boundary of all things. Consequently, the question does not arise of what happened *before* the big bang. The word 'before' implies a temporal order, and this ceases to exist at the singularity. The same is true of the question of causation. It is often asked what caused the creation event. The whole identification of cause and effect with time *order* (cause always preceding effect) is in any case dubious, so that to require the creation to have a cause which precedes it is not in fact necessary.

Moreover, the notion of a preceding causation is clearly meaningless here because temporal considerations cannot be extended beyond the singularity. All of these considerations show that the creation event is physically far more profound in the theory of relativity than it is in the bible. Some of these issues regarding causation and creation will be considered again in chapter 7 in the light of the analysis of time asymmetry in cosmology.

Most cosmologists appear to accept the interpretation just described of the Friedmann cosmological models. What is a matter of dispute though is to what extent these simplified features of the Friedmann models are truly descriptive of the real universe.

It was mentioned that the models described were calculated by neglecting the *pressure* of the universe. Normally if a substance is compressed, a pressure builds up to prevent further compression. It might be expected that in the cosmological case such a pressure would prevent the indefinite shrinkage of the universe as $R = 0$ was approached. In fact, pressure does become very important in the early stages of the expansion. Mostly this is due to the radiation in the universe. Remember that light is red shifted by the expansion. It follows that in the past light was shifted in the opposite direction, i.e. to a higher frequency, and hence a higher energy. But light exerts a pressure proportional to its energy (it is possible to demonstrate the small, but significant, pressure of light by shining a powerful torch on to a 'windmill' delicately suspended in a vacuum, thereby setting it into rotation). The radiation pressure thus builds up and up at progressively earlier stages of the expansion.

In general relativity, pressure is also a source of gravitation. In fact, during the first million years or so of the expansion, the radiation pressure dominated over the matter density in this respect. Because of the gravitating effect of the pressure, the squeezing of the universe cannot be halted. Far from slowing down the shrinkage near $R = 0$, it actually speeds it up. Thus it is not possible to avoid the singularity in the Friedmann models by including the effects of pressure.

Some indication of how seriously to take the possibility of a singularity in the real universe, is provided by a study of models which are more general than Friedmann's. In fact, one of the outstanding mysteries about the observed state of the universe is the reason why it is so uniform anyway. Part of the mystery is due to the fact that every point in the expanding universe is surrounded by a horizon, rather like that around a black hole, which prevents any communication at all between regions which are sufficiently far removed. The horizon which surrounds us at the moment is 10^{28} centimetres away – about 10 billion light years. At that point, the matter is receding from us so fast that, in a sense, it reaches the speed of light (infinite red shift). Beyond that point in space there can be no conceivable communication with us. The horizon grows with time. Eventually matter at a greater distance than 10^{28} centimetres will be visible on Earth.

Conversely, at early times the horizon was very small. In the Friedmann models, at 10^{-18} second after the beginning of expansion it was so small that it encompassed a volume as minute as a single atom. The problem is, if such small regions of the universe were causally disconnected at the start of the expansion, one region of the universe couldn't 'know' what another region was doing. How then did the whole observable universe manage to expand at the same rate?

One answer to this question is that it didn't. A possible scenario is that the universe began expanding chaotically and erratically, and that some dissipative or damping mechanism smoothed it out. This assumption of complete primordial chaos as opposed to complete primordial symmetry has the added attraction that the universe need not have been created in any very *special* condition. If a successful damping mechanism can be found, it would permit a very wide range of initial conditions consistent with the presently observed universe. This issue of how special our universe is will be discussed further in chapter 7.

Several dissipative mechanisms have been suggested. One of these involves the effect of viscosity of neutrinos in a very high density state. Another, much championed by the Russian cosmologists, appeals to particle creation. In section 4.4 it was discussed how particle/anti-particle pairs could be created if an amount of energy $E = 2mc^2$ could be found. This energy can be supplied by tidal effects in the gravitational field. With this mechanism, particle/anti-particle pairs are quite literally created out of bent empty space. The reaction of this creation back on space has the effect of smoothing out the bends. The greater the departure of the geometry of space from Minkowski space, the more prolific the particle production. It follows that in a universe full of chaotic motion, such production effects would tend to smooth things in the direction of their present uniformity. It might even be possible for all of the matter in the universe to have been created in this way rather than at the singularity. Without now invoking the violation of the 'label' conservation laws at the initial singularity, this latter process is predicted to produce as much anti-matter as matter. This would not be a problem if some mechanism could be discovered which would separate the matter from the anti-matter, and prevent most of it annihilating again. For some years the French physicist R. Omnès claimed that such a mechanism exists on the basis of considerations of elementary particle physics. This separation could result in some galaxies being made of matter and others of anti-matter – a perfectly safe cosmic arrangement, because galaxies rarely collide. It would be wise, however, for an ambitious space traveller to establish before he travelled to another galaxy whether it was composed of material the same as himself. Until further theoretical work is done on these topics to indicate how seriously they should be taken (Omnès' work has received much criticism), we seem to have an aesthetic choice between the symmetry of matter and anti-matter with primordial chaos, or the symmetry of a smooth beginning with unbalanced matter.

At first sight it seems that a departure from complete uniformity might prevent the occurrence of the initial singularity altogether. In the Friedmann model, this singularity occurs where all the matter comes together at a point. If the motion was more chaotic, such a concurrence might be avoided. However, an elegant mathematical demonstration has been given by George Ellis and Stephen Hawking that, subject to some very reasonable assumptions about the behaviour of matter at high densities, the existence of at least one singularity in our universe cannot be avoided, even by deviations from exact uniformity. The theorem does not provide much information about the nature of the singularity, or the condition of the universe in its vicinity, except that any particle which strikes it must cease to exist in our space–time. Analysis of certain anisotropic and inhomogeneous models indicate that the past behaviour of the universe as the singularity is approached could be extremely complicated, in contrast to the smooth progress to extinction displayed by the Friedmann models.

Although deviations from uniformity apparently cannot extricate our universe from the pathology of a singularity somewhere in space–time, it can happen that most of the matter in the universe 'misses' it, so that although space–time develops an 'edge', the matter in the universe need not necessarily crash into it. Explosions of this sort, where matter emerges from 'near' the singularity at ultra-high, but not infinite density, have been christened 'whimpers' (in contrast to 'big bangs').

There is still the possibility that the Hawking–Ellis theorem might be violated if the behaviour of matter at ultra-high densities deviates markedly from the general expectation. Few physicists would be prepared to make a strong case for ultra-dense matter violating the conditions of the theorem because these conditions are so reasonable (essentially only requiring that the energy and pressure of matter remain positive). However, although not expected, it is not known whether a negative energy or pressure might perhaps occur at some stage. Actually certain quantum properties of matter do enable negative pressures to arise in some (rather contrived) situations, but singularity-free cosmological models based on this arrangement are a far cry from the real universe.

At a deeper level, quantum effects of space–time (i.e. quantum gravity), rather than quantum matter effects, might prevent the extinction of the universe at a singularity, say, by causing the universe to 'bounce' at a sufficiently high density. As explained in chapter 4, there is as yet no satisfactory theory of quantum gravity, so that this possibility remains purely a conjecture. Tentative attempts to circumvent some of the technical mathematical problems associated with a full theory of quantum gravity by treating a simple model universe directly as a quantum system, thereby checking the 'bounce' conjecture, have led to inconclusive and ambiguous results. In any case, it is uncertain how seriously this sort of quantum cosmology should be taken, because of profound difficulties in the interpretation of quantum mechanics when the entire universe is being considered as the quantum system.

If most of the matter in the universe misses the singularity, or there is a quantum bounce of some sort, the question naturally arises as to what condition the universe was in *before* the bounce occurred. Because of the underlying time symmetry of general relativity, the theory predicts that the large-scale motion of the universe in this previous phase was a reversal of our own phase. It follows that prior to the present expansion the universe would have been in a state of contraction. On a smaller scale, presumably this earlier phase would have had galaxies, stars and cosmologists, though this is really only a conjecture. The wider implications of these models, in which there is no definite creation event, will be considered in the next chapter.

At the time of writing, most cosmologists seem to accept the Friedmann singularity as descriptive of a genuine creation event in the real universe, so it is worthwhile discussing briefly whether space–time can ever get itself into a similar 'creation condition' again. The initial singularity has been compared with that at the centre of a black hole. This is only partly correct. The big bang really represents the time reversal of a black hole. In the former case matter is exploding out of a singularity, in the latter it is collapsing inwards. Nor is this just semantics. The horizon which surrounds all regions in the big bang is the time *reverse* of the horizon round a black hole, so that whereas the singularity at the centre of a black hole cannot be seen by us on the outside, the singularity at the beginning of cosmological expansion is 'naked'. This means that in principle we can look out into the universe and (because of the travel time for light) back in time to see the creation.

In practice, it is not possible to see back before about 10^5 years after the beginning of the expansion, because the cosmological material was opaque to radiation before that time. However, the important principle remains that the universe is a region of space–time which lies in the causal future of a singularity, so that the nature of the universe could not be *predicted* in any way. We cannot say, even in principle, what might come out of a singularity. This is consistent with what is understood about time asymmetry in cosmology, which suggests (see chapter 6) that in some sense, the universe started off in a *random* fashion.

5.4 The big bang

In spite of the qualifications concerning the very early stages of the expansion, the Friedmann models remain the basic working model universes for most cosmologists. Accepting for the moment that $R = 0$ describes a simplified version of a genuine creation event, it is possible to examine in great detail the processes occurring in the Friedmann universe in the early stages of the expansion. Some of the consequences of these processes are observable today, so that this simple model may be confronted with various observational data to test its reasonableness. It turns out that the Friedmann universe does remarkably well, considering its simplicity.

Although it is not possible to continue known physics back as far as the first moment, or even before the onset of quantum gravity at 10^{-43} second, it is possible to construct a model of the universe after the first microsecond or so with some reasonable confidence that the physics is well understood. Charting the course of the universe in these amazingly brief early moments must surely be one of the most awe-inspiring enterprises ever undertaken by science. It is truly incredible that meaningful statements can be made about the condition of the universe when less than one second old.

Like any physical system, the contents of the universe grow hot when compressed, and cool when expanded. The famous red shift of light waves discovered by Hubble can be considered as a cooling of the light due to the cosmological expansion. It follows that during the early stages of the big bang, the universe was very hot indeed because of the enormous compression. For that reason the contents of the universe during this epoch are usually referred to as the *primeval fireball.*

None of the present structure, such as stars or galaxies, which we observe in the universe today could have existed in the fireball.

Even atoms would have been smashed to pieces in the colossal temperatures and pressures. The fireball at very early times should be envisaged as a fluid of all types of elementary particles, strongly interacting with each other, and in a condition of thermal equilibrium.

Some cosmologists have little reservations about discussing the condition of the fireball at *earlier* times than a microsecond, but we take up the story at this moment, when the temperature was about a million million degrees. Although one-millionth of a second may not seem long in human terms, it is a very long time in elementary particle physics. A great deal in the way of interactions between exotic types of particles, some as yet unknown in the laboratory, undoubtedly occurred in that first brief era of violent activity. Much of this elementary particle physics is ill-understood at the moment, but by the end of the first microsecond, all but the most familiar particles would have long since disappeared by disintegration and decay. An infinitesimal moment of glory – the universe filled with billions upon billions of extraordinary particles – and then gone, many of them perhaps never to appear in the universe again!

With the temperature falling rapidly from 10^{12} degrees, the fireball then began the so-called lepton era, with familiar protons, neutrons and electrons as well as muons, neutrinos and electromagnetic radiation (in the form of X-ray photons) all jumbled together in equilibrium. The radiation was so hot that it could create electron–positron pairs. As the temperature dropped, first the muons disappeared, then the positrons. After about 10 seconds the temperature had fallen to a few billion degrees and the principal interest centres on the remaining protons, neutrons and electrons.

At this stage an important new era, called the plasma era began. The temperature is now low enough for the frantically moving neutrons and protons to start combining together to form helium and a few other light nuclei. Detailed calculations indicate that about a quarter of the protons get incorporated into helium nuclei, with a tiny proportion as deuterium and lithium. Thus about 10% of the nuclei which emerge from the fireball are helium, and the rest hydrogen (single protons). This is remarkably close to the present observed abundances for these light elements, which strongly suggests that the primeval fireball be identified with one of the element building factories mentioned in section 5.1. It is also valuable confirmation that the processes which occurred in the plasma era in the real universe were not far from that which the fireball model of the Friedmann universe suggests.

The plasma era continued for about 700 000 years, after which the temperature was down to 4000 degrees absolute (a little cooler than the surface of the sun) and the electrons began to combine with the nuclei to form ordinary atoms. After this had occurred the way was clear for local condensations of matter to form under gravitational attraction. Clumps of gas were whirled up into clusters which slowly contracted to form galaxies and eventually stars and planets.

The temperature of the fireball has been falling ever since because of the continued cosmological expansion. After 10 billion years it is now down to a mere three degrees above absolute zero – less than the temperature of liquid air. It must be counted as one of the great discoveries of science that this feeble, dwindling glow from the primeval fireball was detected by two Americans, Arno Penzias and Robert Wilson, in 1965. It is known as the cosmic background radiation. This fossil of the fiery birth of the universe has travelled more or less unhindered through space ever since the end of the plasma era. It continually strikes the Earth from all parts of the sky. Its existence provides great confidence that the general ideas of the hot big bang model are correct, and that the universe was in a dense, exploding phase about 10 billion years ago.

5.5 Non-standard cosmologies

Until now almost all the discussion given on cosmology in this book has been based on the three types of Friedmann models depicted in fig. 5.5, perhaps with significant deviations therefrom at a very early stage. The reason for this is because it reflects the broad current of opinion among astronomers and cosmologists at the time of writing. However, much of the astronomical observational data are of a tentative and incomplete nature, and wide changes of opinion have occurred in the past. This may happen again.

From time to time other model universes are suggested, which often differ drastically from the canonical big bang model. Many of these alternative cosmologies are based on a modification or even abandonment of Einstein's theory of general relativity, either replacing it with a different theory of gravitation, or a new set of principles altogether. One such alternative, which for many years made a great impact on cosmology, is the steady-state theory. This theory does not have a big bang, and so conflicts with the evidence, such as the existence of the cosmic background radiation, that the universe was in a hot, dense state at some point in the past. The implications of the steady-state theory for the nature of time will be considered in the next chapter.

Among the various non-standard models which do have a big bang, perhaps two are worth a mention: those based on the introduction of an additional cosmic repulsion term in Einstein's equations of general relativity, and models which propose that the universal constant of gravitation, G, varies with time. In section 5.3, it was mentioned that Einstein proposed a modification to the original field equations of general relativity in order to obtain a cosmological model which was static, because Hubble had not at that time discovered that the universe was in a state of expansion. Although the modified equations are perfectly valid contenders for the correct description of gravitation, the presence of the additional term does have some strange and, some cosmologists believe, unsatisfactory consequences. The new term concerned has the physical effect of a universal force of *repulsion* at all points in space. It was by balancing this cosmic repulsion with the gravitational attraction of matter that Einstein was able to construct a static model universe. However, this exact balance is actually unstable, so that a slight perturbation to the equilibrium results either in collapse, or unlimited expansion.

There are many possible homogeneous and isotropic models incorporating the cosmic repulsion which end up with this type of unlimited expansion. Because the repulsion actually *increases* with the separation of two points, the effect of expansion is to increase the repulsion still further. The universe therefore escalates into runaway expansion. The scale factor, R, increases exponentially, so that space grows with ever increasing acceleration, quite unlike the models shown in fig. 5.5 which decelerate as they expand. One such model once proposed by the Dutch astronomer Willem de Sitter (1872–1934) exhibits this type of runaway expansion, but without containing any matter at all – just expanding empty space!

Some of these cosmic repulsion models (though not all) expand from an initial point with $R = 0$, making them candidates for big bang universes. In particular the Eddington–Lemaître model (after Sir Arthur Eddington and Georges Lemaître – a Belgian cleric) displays a very interesting type of behaviour. It starts out by expanding in the same general way as the standard Friedmann models, with decelerating rate caused by the gravitational attraction of the matter. Eventually, the cosmic repulsion overbalances the gravitational attraction, and unlimited exponential expansion takes over. However, during the period when the two effects are about equally balanced, the universe remains nearly static. This 'coasting' phase of the model may be made as long as one pleases by choosing a value for the cosmic repulsion close to Einstein's choice for a static universe. A few years ago the Eddington–Lemaître model was resurrected as a possible explanation for the over-abundance of the strange objects called quasars, observed with high red shifts.

A fundamentally different approach to cosmology was suggested in 1937 by the great British physicist and Nobel laureate, Paul Dirac. Like the astronomer Sir Arthur Eddington, Dirac was impressed by the apparent coincidence that the universe was bigger than an electron by the same enormous factor that electricity is stronger than gravity (both numbers are about 10^{40} for electrons). However, as explained in section 5.5, the size of the universe is continually growing as the horizon recedes, so that this coincidence seems to be due to the accident that we happen to be living at *this* time in the expanding universe rather than any other. In section 7.3 one explanation of this 'accident' will be described. Dirac did not regard this relationship between the two quantities as a coincidence, but proposed instead that it remain true at *all* times. One way in which this is possible is if the force of gravity grows *weaker* as time goes on. The correct amount of weakening is achieved by making G proportional to $1/t$, so as the time t increases, G decreases. This leads to a universe similar to the Friedmann model 2 but expanding twice as fast ('easier' to escape a falling G). Consequently, it is half as old, which poses a time-scale problem, because if the present expansion rate is to be believed it means that Dirac's model predicts that the universe is less than 10 billion years old, whereas evolution measurements indicate that the galaxy has been around for at least 10 billion years. Nevertheless, Dirac's suggestion was developed into a full theory by Pascual Jordan, and later Carl Brans and Robert Dicke. In its latter form, the theory is still a serious contender to Einstein's, and does not pose such difficult time-scale problems.

Starting from a totally different theory, the British astronomer Fred Hoyle and his Indian colleague Jayant Narlikar have also produced a closely similar theory of gravitation with a time-dependent G, in which the mass of a particle is derived from its interaction with distant matter, after the fashion of Mach. Moreover, Dirac himself, even at the age of 75 years, has recently developed his 1937 idea into a full theory of the Brans–Dicke type, reaching different conclusions from his earlier work. All of these theories are subject to falsification by solar-system observations carried out on the motion of the planets, which would be slightly modified by a slowly diminishing gravitational constant. Current observations should, within a few years, reveal any plausible variation of G.

6 The beginning and the end

6.1 The unstable universe

In chapter 3 it was established that the laws of physics do not distinguish a preferred direction of time. The asymmetry in time which is so evident in everyday life, from such diverse phenomena as the behaviour of gases, the dissipation of heat and the propagation of waves, cannot be accounted for by any property intrinsic to the system examined. In all cases the time asymmetry is imposed from without, by virtue of the random formation of *branch systems* – quasi-isolated regions of the universe which separate off from the main environment in a non-equilibrium condition.

Upon close examination most branch systems belong to a hierarchy of some sort. A good illustration is provided by the ice cube example referred to in section 3.1. When a branch system is formed by placing an ice cube in a glass of boiling water, a time asymmetry is established because the low entropy condition of the contents of the glass subsequently pass, with overwhelming probability, to an equilibrium condition of warm water at a uniform temperature. The time asymmetry is imposed on the system by the agency which created it, by executing the formation process at random and not at the outset of an extremely rare fluctuation. The question now remains, how was the disequilibrium achieved which enabled the two components of the glass contents to be formed initially at differing temperatures? The answer might well be 'a refrigerator' in this case. A refrigerator is a branch system which is permanently maintained in disequilibrium by expending energy. The energy supply is *also* a branch system for the following reason. Fuel of some sort (for example, oil or coal) is brought into contact with fire. The system so formed is in violent disequilibrium and rapidly proceeds to equilibrium by combustion. The disequilibrium of the fuel depends on the process by which it was originally formed. In the case of fossil fuels the origin is biological. All biological branch systems are operating in a highly non-equilibrium condition and depend for their continued disequilibrium ultimately on the sun's heat and light. Switch off the sun and all life would soon cease on the Earth.

Wind power and wave power also depend ultimately on the sun's radiation. Their origin is the disequilibrium in the Earth's atmosphere caused by the unequal distribution of radiation falling on the Earth's surface. A moment's reflection reveals that almost all the time asymmetry encountered on the surface of the Earth has this 'sunlight' disequilibrium as its origin. Several examples should be adequate: all biological activity and death, the melting of snow, the wind erosion of rocks, electrical storms, ocean currents.

Some asymmetric phenomena are clearly not ultimately caused by sunlight. The eruption of volcanoes, meteoric impacts and the tides are all caused (at least to some extent) by *gravitational* rearrangements. In addition, various radioactive substances are in obvious disequilibrium. Their origin is of some interest and will now be discussed in a little detail.

In section 5.1 the existence of a cosmic factory to synthesise complex atomic nuclei was suggested. One such factory is the primordial fireball present in the big bang. However, calculations have shown that insufficient heavy elements can be built up this way to account for their observed abundances. As complex nuclei cannot possibly have survived the big bang they must have been created afterwards somewhere. But where? One place already mentioned is the stellar interiors, where element building is responsible for the liberation of energy to supply starlight. It is now generally accepted that massive stars are the major factory for building up heavy elements. Briefly, the scenario goes something like this. Starting with hydrogen, stars gradually synthesise helium during their quiescent stable phase (our sun is currently in the middle of this phase). Eventually the helium gets hot enough to build heavier elements (principally carbon) and in turn these elements build still heavier elements and so on, in a long chain. The details of the chain are extremely complex and depend upon the subtleties of nuclear physics. Eventually, however, these stars will contain a certain small proportion of heavy elements. The heaviest ones (such as uranium) actually represent an energy *loss* to the star, because the nuclei of these elements liberate energy if they undergo fission (splitting), unlike the light elements which release energy during the opposite fusion process.

The problem remains of how to get these heavy elements out of the stars and into the surrounding regions of the galaxy. A spectacular mechanism for this is provided by supernova explosions – fantastic cataclysms which blow most of the star to pieces in an energy outburst millions of times greater than the steady rate of starlight. These events are, thankfully, very rare. One thought to have occurred in our galaxy was recorded by Chinese astronomers in the year 1054 as a star as brilliant as the planet Venus. After a few days the extreme luminosity faded. What is left of this star now is an irregular shaped object called the Crab nebula, consisting of a mass of gas streaking out at high speed from a small object near the centre, thought to be a neutron star.

Having spewed its heavy elements out into the galaxy in this fashion, the dead star has provided tenuous material for the formation of a new generation of stars rich in these new elements. The attendant planetary systems of the second generation would then contain a high share of these elements. Our sun is only about one half as old as the universe or less, so there has been plenty of time for such supernovae events – rare though they may be – to have seeded the galaxy with all the heavy elements around us. It is a remarkable thought that the carbon (which is the basis of all terrestrial life) in our bodies represents the debris from the violent death, aeons ago, of a former generation of stars.

According to this picture of element formation and distribution, the radioactive substances on the Earth, the disequilibrium of which liberates some of the energy used to supply our electricity, owe this disequilibrium to the conditions in the interior of long dead stars. So it is that one way or another, most of the disequilibrium around us which enables our environment to change with time owes its existence to the formation and evolution of the sun and stars. This pattern of disequilibrium through starlight and nucleosynthesis is repeated throughout the universe.

The whole cosmos is in an unstable state, with vast chasms of cold emptiness punctuated sporadically by white hot stars. These tremendous powerhouses of energy are continually pouring out starlight in an attempt to redress the balance, and restore thermal equilibrium.

The thermodynamic disequilibrium of the universe was noted in connection with Olbers' paradox in section 5.1. But this paradox is resolved only to produce another. For how did the universe *get* so unstable in the first place? This is the same question encountered with the branch systems, only this time one cannot appeal to outside interference for the answer because one is dealing with the whole universe, which has no 'outside' to do the interfering.

Naturally, the problem could be dismissed by saying that the universe was simply made that way in the big bang, with inbuilt disequilibrium right from the start. This retort has two rather unsatisfactory features, one philosophical, the other physical. Science is supposed to *explain* features of our environment. It does not constitute an explanation to say that things are what they are because they were what they were. Secondly, there is good evidence, such as the cosmic thermal background radiation, that the universe was in thermal equilibrium some time in the past. If this equilibrium was a reality, how has the present disequilibrium been achieved? How can a stable universe become unstable? Equilibrium is associated with maximum entropy or disorder. How can a chaotic, disordered universe turn into a structured, ordered universe, when all our experience teaches that order gives way to disorder and not vice versa?

To answer this question it is necessary to return once again to the very early stages of the expansion and examine carefully some of the processes occurring in the primordial fireball.

Firstly, the nature of the cosmic instability must be understood in more detail. This instability (at least to a great extent) is the production of starlight. The energy radiated is supplied by the synthesis of the nuclei of heavy elements in the stellar interiors. When two light nuclei fuse, part of the total mass is converted to radiation energy (according to Einstein's formula $E = mc^2$) which then percolates slowly through the outer layers of the star and off into space. This process represents an entropy increase because the energy which was locked up in the nuclei gets spread out into space, a process which may be envisaged as an increase in disorder. The fused nuclei, by ridding themselves of some energy, are more stable as a result.

This principle of achieving stability by the shedding of energy is really quite general. A ball placed on top of a hill (see fig. 6.1) is unstable. A slight movement will send it rolling down the slope, during which time it will gain kinetic energy (energy of motion) at the expense of its gravitational potential energy (the energy required to carry it to the top of the hill). Its gathering speed will cause the ball to overshoot the valley and climb the hill the opposite side, but some of its energy will be lost by friction with the ground and air resistance, and get converted into heat, which will flow away into the surroundings (thereby increasing the entropy) in accordance with the second law of thermodynamics. Thus, bit by bit, as the ball rolls back and forth, its energy will be dissipated away and it will come to rest at the bottom of the

Fig. 6.1. Instability. The ball at the top of the hill is in an unstable condition. A slight disturbance will send it rolling under gravity down into the valley. Eventually it sheds its energy (through friction) and comes to rest in stable equilibrium at the bottom of the valley.

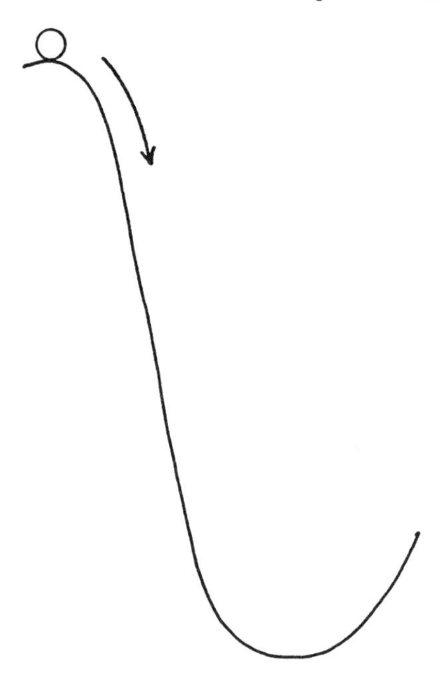

valley. The energy which it possessed at the top of the hill will have been shed as the price for achieving its stable equilibrium in the valley below. The organised activity of the ball has been converted into the disorganised activity (heat) of atomic motions, in accordance with the law of entropy increase.

The ball example can also be used to illustrate nicely the concept of *metastability*. In fig. 6.2 the hill has been drawn with a small hollow in the top. A ball placed in this hollow will be in a type of local equilibrium, because slight perturbations will not knock the ball out of the hollow; if it is pushed a little to one side, it will merely roll back again. However, if it is given a big push it will climb up over the barrier and roll down the hill. The ball in the hollow is said to be in a *metastable* state.

Now replace the ball by the nucleus of a light atom, say, hydrogen, for the sake of definiteness, and the Earth's gravity

Fig. 6.2. Metastability. The ball is protected from the incline by a small barrier. Although in equilibrium, it is only metastable, because a sufficiently large push will send it over the barrier and into the valley below. This is close to the situation in nuclear physics, where the barrier is the electric repulsion between protons and the valley is the strong, but short range, attraction of a nucleus. If a proton hits the nucleus fast enough, it will surmount (or tunnel through) the electric barrier and drop into the nucleus, shedding its energy in the form of γ-rays.

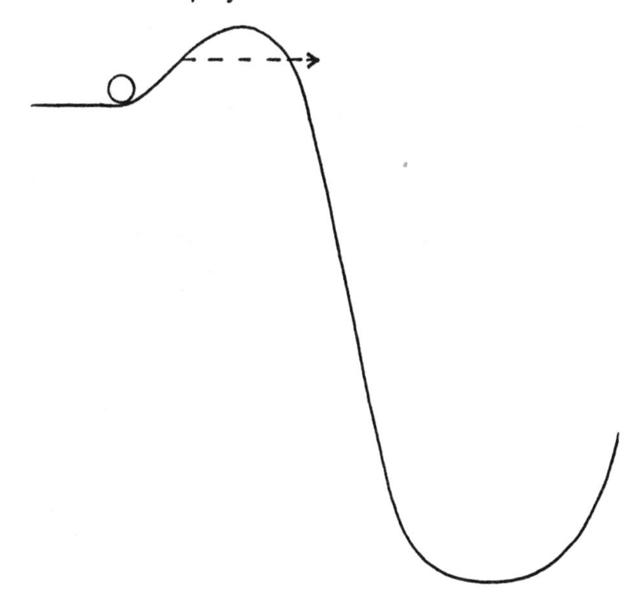

(which pulls the ball down the hill) by the strong force of attraction of another nucleus of some sort. The proton (hydrogen nucleus) wants to 'roll down' into the attracting nucleus ('the valley') and would readily do so if it wasn't for the fact that it is electrically charged. This charge on the proton is strongly repelled by the similar charge on all the protons in the other nucleus. Although the nuclear attractive force is even stronger, its range is very short, and the proton has great difficulty in getting near enough to come within range and so drop into the nucleus. The proton is therefore in a *metastable* state, prevented from falling into the nucleus by an electric barrier. If it is given a small push, it will bounce back off the barrier again. But a very large push will send it over the top and into the nucleus, where it will 'fuse', emitting energy (in the form of γ-rays) as a result. The proton bound in the nucleus ends up in a stable state at the bottom of the nuclear 'valley'.

In the terrestrial environment hydrogen atoms have nowhere near enough energy from their thermal motion to overcome the electric barrier. Thus, although hydrogen is, strictly speaking, metastable, it is only at enormous temperatures that the protons possess sufficient energy to undergo nuclear fusion. In fact, the situation is helped a little by quantum mechanics because it is possible for the proton to do a disappearing trick for a very short time (just as virtual protons can appear out of nowhere for a brief moment). This time is, however, long enough for the 'missing' proton to travel a little way and reappear on the other side of the barrier, a process known, for obvious reasons, as the tunnel effect. When the tunnel effect is taken into account, it may be calculated that a temperature of several million degrees is required to enable protons to fuse together in a nucleus of helium. (Actually this is not a single process but several. Two neutrons are also required to participate.) We may deduce that the central temperatures of stars like the sun, now in their hydrogen-burning phase, are several million degrees.

Having established that a large part of the instability in the universe is really the metastability of hydrogen it is necessary to understand why the universe is made mainly of hydrogen.

Assuming that the basic outline of the hot big bang model is correct, then the condition of matter in the early stages must have been one of individual constituents of matter all moving about independently at relativistic speeds. The colossal temperature would have smashed all nuclei, and their constituents also, into as many pieces as they possess. The thermodynamic condition of this primordial fireball would have been one of local equilibrium, for although the universe was rapidly expanding and cooling, the matter was so dense that it almost instantly readjusted itself to the changing conditions. However, after a few hundred seconds, the temperature had fallen low enough for the separately moving protons and neutrons to fuse together to form complex nuclei, without these being instantly disintegrated by the intense radiation. Some nucleosynthesis did indeed occur and, as mentioned in section 5.4, calculations indicate that about a quarter of the protons ended up in helium nuclei, with a very small additional fraction in other elements, such as deuterium and lithium. But the presence of the electric barrier has a crucial effect here. For only during a limited period of time was the temperature of the plasma both *low* enough to prevent disintegration of the synthesised helium nuclei, yet *high* enough to enable the protons to overcome the electric barrier. After a while the synthesis was simply 'switched off' by the barrier, and the protons were frozen in a metastable state, eventually to become hydrogen atoms.

The mode of transition of the universe from equilibrium to disequilibrium is now apparent. Had the universe remained a fireball there would be no time asymmetry of the sort encountered in the actual world at present. But the changing conditions in the expanding fireball modified the equilibrium form of matter from separately moving individual particles to heavy nuclei. The disequilibrium of the universe is therefore to be sought in its *expansion*. It is this very expansion which alleviates the problem of Olbers' paradox discussed in chapter 5.

6.2 The absorber in the future

Leaving aside for the moment the implications of cosmology in general, and the big bang in particular for thermodynamic time asymmetry, let us focus attention on the other asymmetric process discussed in chapter 3, that of retarded wave motion.

In 1945 John Wheeler and Richard Feynman, two of the most distinguished American theoretical physicists of the post-war years, published a novel and elegantly simple explanation of why electromagnetic waves only travel outwards into the universe or, more picturesquely, why radio signals only travel forwards rather than backwards in time. Curiously, the motivation behind the Wheeler–Feynman theory was not one directly connected with time asymmetry but rather with the structure of charged elementary particles. These authors sought to remove certain formidable mathematical difficulties which for decades have plagued the descriptions of the interaction of charged particles with the electromagnetic field. Although the new theory meets some success in this direction, all attempts to produce a quantum mechanical version of it seem inevitably to reintroduce the mathematical difficulties once again. For this reason, much of the original advantage of the theory is lost, and the arguments for accepting it are less compelling.

Nevertheless, Wheeler and Feynman's ingenious idea has proved an attractive framework for all sorts of speculation on the subject of time asymmetry and cosmology, particularly among cosmologists themselves. Longstanding and tenacious proponents of the Wheeler–Feynman theory are Fred Hoyle and Jayant Narlikar, who have extended the original conception to a theory of gravitation (discussed briefly in section 5.5) and even of elementary particles. We shall not dwell upon these wider issues here, but restrict the discussion to an outline of Wheeler and Feynman's original idea itself.

Recall that Maxwell was the first person to combine together the known laws of electricity and magnetism into a unified electromagnetic field theory which predicted the existence of electromagnetic waves. The mechanism for the production of these waves is an electric current, now known to be due to the motion of electrically charged particles, such as electrons. In order to become a source of waves, a charged particle must be *accelerated.* The electromagnetic field which surrounds the particle is forced to adjust itself to the changing motion of the particle, and the resulting disturbance propagates away in the form of waves. These waves carry energy so that the accelerating particle can be said to *radiate*, or emit radiation. The radiated energy must be paid for, and it appears to be at the expense of the energy of the particle, the accelerated motion of which is therefore damped. The effect of the damping is to exert a force on the particle, called the radiation damping force. In practice this force is extremely small. The branch of physics which deals with the interaction of electromagnetic fields and moving charged particles is called electrodynamics.

Because electrodynamics based on Maxwell's theory is completely symmetric in time, the possibility also arises of the reverse process, where electromagnetic waves strike an accelerating charged particle and become absorbed as a consequence. This is a well-known phenomenon also. What is at issue here is not the correctness of the reversibility of electrodynamics, but the following: a particle which is arbitrarily accelerated apparently 'causes' the radiation of retarded waves in a coherent pattern of motion which spreads outwards from the vicinity of the particle. It does not 'cause' the reverse process where waves which, having travelled inwards from the remote parts of the universe in different directions, impinge upon the particle in a *coherent* wave pattern, and become absorbed. In short, organised waves get emitted, only disorganised waves get absorbed. One way of expressing this is to say that accelerating a charged particle causes waves to be emitted into the future, but not into the past. (A wave 'emitted into the past' is a use of backwards-time terminology to describe a wave coming from the past and being absorbed. Not only is it easier to use this terminology, but also it avoids introducing a language prejudice to mask the underlying time symmetry of what follows.)

Wheeler and Feynman did not change the basic form of Maxwell's theory, but found a (possible) deeper reason for why only future-directed radiation occurs, rather than the dismissal that the universe is simply 'made that way'. This they did by analysing what would happen if an accelerating charged particle emitted radiation *equally* into the past and future. Of course, sending signals into the past involves all sorts of paradoxes similar to those encountered in connection with tachyons mentioned in section 2.5. Clearly, this type of behaviour by a single, charged particle is in straight contradiction with experience. Nevertheless, Wheeler and Feynman guessed that there might be a *collective* motion by many similar particles, the waves from which, when taken all together, might be the familiar and acceptable fully-retarded (future or outward-going) form, even though *individually* they were time symmetric.

How can this come about? The underlying mechanism is the well-known phenomenon of *interference*. A good way of demonstrating interference effects is with water waves. Drop two stones close together on to the surface of a calm pond. The ensuing wave patterns from each stone do not propagate independently, but interfere with each other to produce a criss-cross network of local peaks and troughs. Where the peaks and troughs from the separate wave patterns happen to coincide, they reinforce each other, but where the peaks of one meet the troughs of the other a *cancellation* occurs, and the water surface remains relatively undisturbed.

Interference effects with light waves are also familiar. The areas of colour seen when a thin film of oil rests on a reflecting surface, such as a puddle of water, are due to the cancellation by interference of certain wavelengths (colours) of the incident white light, and the enhancement of other wavelengths.

Wheeler and Feynman found the following remarkable result. Suppose a single, charged particle in empty space, when set into motion, radiates symmetrically one half advanced waves into the past and one half retarded waves into the future. Then that same particle, when placed inside an *opaque box*, will only radiate fully retarded waves into the future. Open the box, and the advanced waves will reappear!

What happens inside the box is this. The waves from the accelerating particle move outwards until they strike the inside surface of the box, where they set into motion the charged electrons from the box atoms. The retarded wave strikes the box a little after it leaves the vicinity of the particle, but the advanced wave strikes the box *before* the particle has even been moved! So, paradoxically, the electrons in the box vibrate in *anticipation* of the subsequent motion of the charged particle. The prospect of causing a response in the walls of the box at an earlier time may appear a little bizarre, because in human experience a cause always precedes its effect. Nevertheless, in physics the difference between cause and effect is not really relevant; all that matters is interaction. It is quite permissible either to interchange cause and effect, or else to have cause following effect in time, provided that everything is self-consistent.

The vibration of the box electrons (both before and after the motion of the original charged particle) will themselves generate waves and, according to Wheeler and Feynman's assumption, these waves will also be radiated into both past and future. Thus the original moving particle, when placed inside the box, thereby becomes associated with a complex pattern of advanced and retarded waves from the walls of the box. These waves will all interfere with each other in a very complicated fashion. The salient feature of Wheeler and Feynman's work was to demonstrate by a simple calculation that, provided the box is fully opaque (so that no waves whatever can penetrate to the exterior) then the advanced waves from the *box* just cancel the advanced waves from the *source* particle. Moreover, they also enhance the retarded waves from the source particle up to full strength. The effect of the response waves from the box on all the charged particles is to *cancel* all the anticipatory motions occurring before the source particle is moved, and to produce exactly the right radiative damping force on the source particle to account for the transfer of energy from the particle to the walls of the box. To an inhabitant inside the box, the electrodynamic behaviour of the system is thus entirely in accordance with our everyday experience. If, however, the box is not fully opaque, paradoxical advanced effects still occur.

Having presented this demonstration of producing fully retarded waves from time-symmetric waves by using the response of a box, Wheeler and Feynman gave an explanation of where the time asymmetry was inserted. After all, if the entire system is supposed to behave symmetrically in time, then the reverse pattern of waves could be used equally well to produce a fully advanced wave from the source. The key to the asymmetry lies in the mechanism of *absorption.* If the box is opaque, the waves striking the interior surface must be absorbed, which in practice means that they are converted into heat. The vibrating electrons collide with the atoms in the wall and set them into thermal motion. The generated heat then *dissipates away* through the walls in accordance with the second law of thermodynamics. For the reverse effect to occur, and advanced waves to be produced, vast numbers of atoms in the box would have to collide in an appropriate way. For they must transfer their thermal motion to the electrons at precisely the right moment to cause them to radiate, collectively, a coherent wave on to the source particle inside the box. According to the principles outlined in chapter 3, such a situation, though not impossible, is overwhelmingly improbable.

By invoking a mechanism of absorber response, Wheeler and Feynman were thus able to place the origin of the time asymmetry of electromagnetic radiation squarely on the shoulders of thermodynamics. A fundamental way of appreciating how this has been achieved is to observe that if the box is completely opaque, so that no radiation flows away into the space beyond, then the action of the fields inside the box can be replaced by a direct action-at-a-distance between the charged particles. This direct action of particle on particle is not of the instantaneous type which characterises Newton's theory of gravitation, but is a *delayed* action which propagates at light speed. It is, moreover, an action which operates both forwards and backwards in time. As we have seen, such an action principle, though a little unusual, is indistinguishable inside an opaque box from the results of Maxwell's theory, based on the propagation of disturbances through a field. However, the advantage of describing electrodynamics solely in terms of particle interactions is that it places the time asymmetry back in the subject of many particle motions, or thermodynamics, where it is well understood. According to this theory, it is no longer necessary to account for the time asymmetry of waves in the electromagnetic field, for there *is* no field at all.

Naturally the absorber theory of Wheeler and Feynman can only be taken seriously if the real world behaves like the interior of a perfectly opaque enclosure, otherwise we should have to contend with unpleasant advanced effects ('backwards causality'). There is certainly nothing very opaque about the universe in the vicinity of our galaxy. Indeed, light may travel for many hundreds of millions of years without encountering a substantial amount of matter. Whether or not all radiation will eventually be absorbed rather depends on the condition of the universe in the far future. An insistence on the absorber theory therefore amounts to a statement about how the universe will end, so that in a certain sense the local behaviour of electromagnetic radiation enables us to look into the future and foretell what will happen to the cosmos. A simple calculation then shows that, within the standard context of the Friedmann cosmological models, the universe is going to collapse.

6.3 The death of the universe

It is a curious commentary on the impact of time asymmetry on the human intellect that most people expect there to have been a past moment when all things were created, but rarely entertain the thought that there may be a future moment when all things will end. But of course, from the standpoint of physics, all evolution is reversible, and the question of the universe coming to an end is reduced to a matter of deciding whether the large-scale motion of the universe is such as to bring about a reversal of its present development.

Before discussing the nature of its demise, the conditions necessary to bring the universe to such a catastrophe will be described. Referring back to section 5.2, and in particular, fig. 5.5, it may be seen that two alternative futures exist on the basis of the Friedmann models. In models 1 and 2 the universe continues to expand for ever, but in model 3 this expansion is stopped at some stage, and then reversed. Taking the Friedmann models at face value, this 'recontraction' ends up by squeezing the universe to extinction at a final singularity, identical to the one at the start of the expansion. Thus the model which is finite in space is also finite in time, and indeed, time-symmetric. The condition necessary to initiate the onset of collapse is in fact the analogue of the Schwarzschild radius criterion for black holes.

Provided the density of the universe is high enough, collapse cannot be avoided.

At the present epoch the critical mass density necessary to produce a recontracting universe is about 10^{-29} gram per cubic centimetre or an average of about one atom per 100 litres of space throughout the universe. It is known that the density of luminous matter (all the stars and so forth) is only about 1% of this. The issue therefore hinges on how much matter, or energy, there exists in other forms. For example, the intergalactic spaces could contain a vast quantity of material, or the galaxies could contain a large number of very dim stars or black holes. In addition, the universe might be filled with a huge quantity of gravitational waves, or neutrinos, both of which interact so feebly with matter that such a background would be exceedingly unobtrusive.

The determination of the energy contribution from these sources receives a great deal of attention from astronomers. The technical problems and complicating factors are manifold, so that frequent changes of opinion occur both for observational and philosophical reasons. The early part of the 1970s has seen a general movement of opinion towards a high density universe, which movement has recently been sharply reversed.

The issue does not hinge on the density measurements alone. The rate of deceleration of the expansion can be directly measured from the pattern of motion of the galaxies themselves (recall that looking far out into the universe gives a view of what the expansion *was* like in the remote past, enabling the amount of deceleration to be estimated). In practice this measurement tends to overestimate the deceleration because of slow changes in the brightness of galaxies. It seems safest to say that the verdict of the recontraction debate is still open.

In the previous section the Wheeler–Feynman theory was discussed, and the condition for its validity mentioned. This is the complete opaqueness of the universe, a condition depending on the cosmological motion in the far future. It turns out that all the ever-expanding Friedmann models are inconsistent with this opaqueness requirement. The recontracting model is, however, perfectly opaque to radiation. The current evidence in favour of a low density, ever-expanding universe should therefore be considered as evidence against the absorber theory.

The events which would befall the universe if it began to collapse would be a return to the fireball conditions of the big bang. This return would come about very gradually, taking many billions of years. For the greater part of the recontraction, the large-scale aspects of the universe would remain largely unchanged because of the delay in the light travelling from distant regions. Eventually, however, a general implosive pattern of motion would be discernible as the galaxies slowly fell on each other and collided. The temperature of the thermal background radiation, augmented by starlight, would slowly rise throughout the contraction, and in the later stages would be hot enough to vaporise the stars. The remorseless cremation of everything would then begin apace as the ever-quickening matter-crushing began. The final fireball would pass backwards through the sequence of states already discussed in connection with the initial big bang, ending up with the whole universe falling together into a space–time singularity. Gravitation is both midwife and undertaker of the universe.

As far as the Friedmann models are concerned, the alternative to crushing and fiery extinction is a frozen waste. If the universe continues to expand for ever then full thermodynamic equilibrium can never be properly achieved. Nevertheless, the present extreme disequilibrium upon which our lives depend, with the cold vastness of space punctuated irregularly with white hot stars, cannot last for ever. Eventually, all the nuclear fuel will be exhausted and the stars will go out. One by one they will either blow themselves to pieces as supernovae, or else slowlv dim and cool. Many will perhaps collapse into black holes The whole process may take billions of years to come about, but it is assured in the long run.

As the expansion continues, so the fading galaxies will gradually disperse and become invisible. Any matter in them which is not eaten by black holes will slowly cool to reach the ever-dwindling background temperature of space. Little of consequence will then happen in this cold, dark, empty wilderness. Occasionally, a sudden catastrophe, such as the collision of two neutron stars, or black holes, will restore momentary activity in a burst of gravitational radiation, and even, one supposes, those unimaginably rare thermodynamic fluctuations will occur from time to time to light up remote corners of the darkness. Otherwise all will be ended.

There can surely be few predictions in science so profoundly depressing as this living death for the universe.

6.4 Worlds without end

A radically different approach to the subject of the time evolution of the universe has been proposed by a number of cosmologists. In 1946 two British astrophysicists Hermann Bondi and Thomas Gold speculated that if the universe appeared the same (on the large scale) from place to place, perhaps it also stayed the same from time to time. In that case the universe as a whole would not really change at all. Naturally, it must continue to expand, and to expand at the same rate always. Normally, the expansion would reduce the density of galaxies in the universe, so to make the expansion consistent with the proposed lack of time evolution, Bondi and Gold postulated that new galaxies are constantly being formed to fill the 'gaps' left by the recession of the existing ones. The material which goes to make up these galaxies is *continually created* as the universe expands. There is no big bang creation in this model. Matter enters the universe all the time. The overall behaviour of this unchanging universe is therefore not static, but a steady-state, with individual stars and galaxies evolving through their life cycles and decaying, and newly created matter always forming up to boost the population of young stars. Thus, if the universe were in a steady-state, it would have no beginning and no end.

An obviously questionable feature of the steady-state theory is the mechanism whereby matter can enter the universe continually. Quantum particle creation from the gravitational field is negligible cosmologically under the present conditions, so some new principle is required. Such a principle was devised by Fred Hoyle, in the form of a new type of field, called the creation, or C field. This C field was endowed with negative energy so that when coupled to matter the creation of an atom (taken for simplicity to be hydrogen) would be accompanied by the enhancement of the negative energy field. The total energy would thereby be conserved, so the whole process would be consistent with the general theory of relativity. It would not, of course, be consistent with the various 'label' conservation laws of elementary particle physics. However, the required creation rate is really very low; about one atom per year in a volume the size of a laboratory – quite undetectable in practice, but sufficient to compensate for the falling density caused by the cosmological expansion.

In its later stages, the theory was developed in great detail by Fred Hoyle and Jayant Narlikar, and for a number of years enjoyed great popularity. Towards the mid 1960s the discovery, first of unmistakable evolutionary effects in the universe, and then finally of the thermal background radiation, gave strong support to the belief that the universe was in a hot dense state a few billion years ago, and could not therefore be in a steady-state. So this imaginative, and somewhat controversial theory, has now been generally abandoned.

Nevertheless, the philosophical lure of a universe without birth and death can be very compelling. A way in which these philosophical advantages of the steady-state theory can be procured, without abandoning the successes of the big bang model, is achieved by the so-called oscillating universe. This cosmology is based on the Friedmann recontracting model but with the additional assumption that the universe survives its encounter with the singularities at both temporal extremities (a number of ways in which this might occur were mentioned in section 5.3). If this were so, at the end of its cycle of expansion and recontraction, the universe would reach a very high density state and then 'bounce' out into a subsequent cycle of expansion and recontraction similar to the first. Continuing this process indefinitely, the cosmological motion would thus consist of an unending series of oscillations between a maximum and minimum size (see fig. 6.3). Such a universe would, like the steady-state theory, be without beginning or end. However, the violence experienced in the high density phases would destroy all the structure and information of the preceding phase, so that the evolution of each cycle could begin afresh.

It might be wondered how the oscillating universe avoids reaching thermodynamic equilibrium. An analogous laboratory system is depicted in fig. 6.4, and consists of a gas confined in a cylinder beneath the weight of a piston. If the piston is depressed firmly from the equilibrium position and then released, it will spring upwards sharply because of the increased pressure from the compressed gas inside the cylinder. The inertia of the piston will cause it to overshoot the equilibrium position, thus rarefying the gas. It therefore falls back again, overshoots once more and compresses the gas. This cycle of expansion and recontraction continues, in analogy with the motion of the oscillating universe. For a number of reasons the laboratory system will not continue to oscillate for ever. With each upward motion of the piston the gas in the cylinder responds by an expansion which always lags a little behind the piston motion. This sluggishness of the gas means that it is always a little out of equilibrium with the apparatus. It follows that every cycle, the entropy of the gas goes up slightly as it tries to restore equilibrium. This entropy increase appears as a temperature increase of the gas, which

Fig. 6.3. New-deal universe. If the recontracting universe survived its encounter with the singularity (somehow) it would bounce on for ever. Cosmic 'friction' effects would generate entropy in the form of heat, which would rise without limit in cycle after cycle. To avoid this, it has been suggested that the universe is 'reprocessed' at the start of each new cycle, providing a 'new deal' in which not only the entropy, but all of physics also, may change. Most new deals would be a poor deal for life, which could only evolve cosmologists in cycles much like our own.

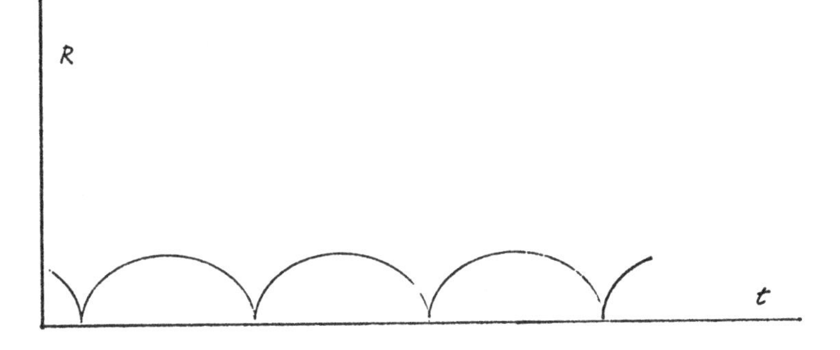

heats up at the expense of the energy in the piston. Consequently, the piston motion is gradually damped, and the system eventually comes to rest with the gas somewhat hotter and the entropy greater. If the piston is moved by a machine, the oscillations will continue until the external energy supply of the machine fails, then the motion will eventually come to rest.

In the cosmological case the entropy generated by the lagging expansion described above is exceedingly small. Much more entropy is produced by starlight as explained in section 6.1. However, the general principle is the same, and at first sight it seems that as the entropy of the universe continues to rise, so the cosmological motion will be slowed to a halt. However, this clearly cannot be so, because a self-gravitating universe can never be in equilibrium – it will inevitably collapse onto itself. The gravitational field has the effect of an unlimited energy supply, and the oscillations apparently continue indefinitely. In fact, calculations show that they actually *grow* in amplitude.

If the entropy of the universe rises from cycle to cycle we manifestly do not live in an oscillating universe of the type

Fig. 6.4. Analogue of the bouncing universe. If the piston is depressed and released, it will bounce up and down, expanding and compressing the gas in a series of oscillations like the universe shown in fig 6.3. However, unless the piston is driven by a motor, these oscillations will eventually cease, as the organised motion of the piston gets converted into the disorganised motion (heat) of the gas, in accordance with the law of entropy increase. When the piston stops, the entropy can rise no further. In the cosmological case, this does not happen. The oscillations are driven for ever by gravity, and the entropy (heat) rises without limit.

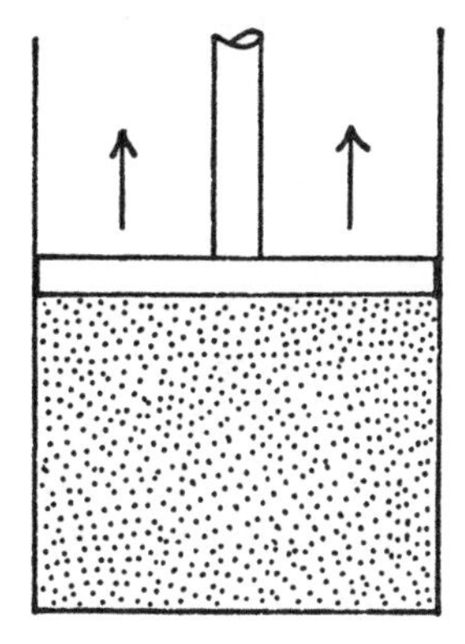

described. By far the greatest amount of entropy in the cosmos is in the form of the background radiation, which is very low. In fact, the entropy produced by the emission of starlight in just one previous cycle would account for all this background radiation. So at most we live in the *second* cycle of oscillation.

Some authors have suggested that the entropy doesn't survive the high density phase between cycles, each new cycle being started off 'reprocessed'. If the laws of thermodynamics are violated at the end points of the cycle, so might any of the laws of physics (though curiously general relativity, the least well established of all these laws, is always assumed to remain inviolable). The extreme limit of this philosophy is to assume that all the laws, and maybe even the naturally-occurring constants, such as the charge on the electron or Planck's constant (after Max Planck, German physicist, 1858–1947), are reprocessed for each cycle. If nothing survives from one cycle to the next, they are really physically disjoint universes, and we may as well regard them as an infinite ensemble all existing simultaneously. In some of them the laws and constants would be similar to our own, and the formation of life would occur. Most of them would be wildly different, with life made impossible. The reason why we observe the sort of universe that we do is then made the responsibility of biology. Only a small subset of all possible models can give rise to cosmologists to wonder about them. The curious inversion of using biology to explain physics, and even cosmology, may appear intriguing to the reader, who will encounter more of this sort of speculation in chapter 7. However, the idea has a philosophical basis only, and is not a physical theory. It cannot be falsified by experiment or observation.

6.5 Order and disorder in the universe

In section 6.1 the origin of the disequilibrium of the universe was traced to the sudden change in the equilibrium form of matter when the primordial fireball expanded and cooled. Essentially, it is because we live in a *moving* cosmos that it has become unstable. In a sense, the global motion of the universe can be thought of as a kind of 'outside' interference with the local thermodynamic systems of matter and radiation. In chapter 3 it was carefully explained that such interference, while necessary for the production of time asymmetry, was not in itself sufficient to determine the direction of the asymmetry in time. An additional assumption of random microscopic motions – molecular chaos as it is sometimes called – is also necessary. If we regard the universe as a whole like a gigantic branch system, there is now an intriguing problem to try and find a *cosmic* explanation of molecular chaos.

One explanation has already been disposed of. This was Boltzmann's suggestion that we are living near the bottom of a gigantic cosmic fluctuation from equilibrium, where the *Stossahlansatz* is extremely likely to be correct. This is inconsistent with what is known about the past history of the expanding universe.

Another possibility (perhaps no less fanciful, though much more vigorously promulgated) is that there is not just one universe, but a whole ensemble, maybe an infinity of them. This could come about either by arranging for successive cycles of oscillation, as discussed in the previous section, or at a quantum mechanical level where, according to a heretical interpretation of the theory, all the possible alternative quantum worlds co-exist in a stupendous array of parallel universes! Whatever the conjecture, an ensemble of universes would enable all possible initial patterns of microscopic motions to be realised. The observed universe is then simply a *typical* member of the ensemble, chosen *at random* from the whole collection. Only an infinitesimal fraction would show 'miraculous' behaviour (such as gases unmixing).

Many people would be rather suspicious of the idea of an ensemble of universes, and would prefer simply to say that the reason the observed universe started off with random microscopic motions is because it happened to be made that way. Whatever point of view is taken, it is clear that a time-asymmetric universe does not demand any very *special* initial conditions. It seems to imply a creation which is of a very general and random character at the microscopic level. This initial randomness is precisely what one would expect to emerge from a singularity which, as mentioned in section 5.3, is completely unpredictable.

One consequence of the assumption that the initial microscopic motions are chaotic is that influences which reach the Earth from different directions of the sky are quite independent and uncorrelated. For example, electromagnetic waves are continually bombarding the Earth, in the form of starlight, X-rays, γ-rays and, most significant of all, the low temperature thermal background radiation which is apparently left over from the big bang itself. Precisely because this radiation is thermal it carries no detailed information about the primordial fireball. There are no coherent messages reaching the Earth in contracting spherical waves. That is to say, advanced electromagnetic radiation is not present because the very special microscopic collaboration necessary to set up converging patterns of waves is ruled out by the assumption of initial random wave motions. (If, alternatively, the Wheeler–Feynman absorber theory is believed, then this step is not necessary of course. The fully-retarded nature of radiation is supplied instead by the thermodynamic properties of the absorbing matter.)

A totally different conception of order and disorder in the universe has been discussed by a number of cosmologists, notably Thomas Gold and John Wheeler.

Suppose that among the primordial chaos there exists a *plan*, a subtle pattern of motions which, although not significant at the time of the big bang, contains the seeds of future miracles. Perhaps the fireball birth of the universe only *appears* to contain random microscopic motions to our eyes, which cannot distinguish hidden cooperation between countless numbers of particles, all moving in different ways and on circuitous routes to a common, coherent end. Could there exist in the universe a hidden order 'folded-up' among apparently randomly moving constituents?

A clever illustration of how we might not spot order 'folded-up' in apparent chaos has been invented by the British physicist David Bohm. Take a jar of treacle and a mechanical spoon to stir it up with. Place a spot of black dye somewhere in the treacle and start to stir. The dye will be drawn out into a thin thread by the motion of the spoon. After a large number of turns the thread will be so thin and so convoluted that the jar of treacle will seem at a glance to be a uniform grey mixture. However, the pattern of dye folded up in the treacle is by no means random. Although the structure of the thread is very complicated, the order is simply hidden – it has not disappeared. This fact is dramatically illustrated by winding the spoon backwards in the reverse direction. The thread actually untangles itself and is slowly drawn backwards into a blob again. The order has reappeared! Maybe the universe is like that; folded-up order among chaos, with the order unfolding at some future time?

This is just the situation which Gold has envisaged. At this epoch of the universe, nothing very miraculous is happening. Branch systems form at random and increase their entropy in the usual way. Disorder rather than order appears in the universe – cars rust, people die, ice melts. New cars, people and ice replace them, but only at the expense of energy dissipation and entropy increase in the wider universe. Total disorder always increases. But what if at some future date this pattern of asymmetry were to reverse? Such a bizarre situation is certainly possible. Although the vast majority of microscopic motions in the big bang give rise to purely entropy-increasing worlds, a very, very special set of motions could indeed result in an initial entropy increase, followed by a subsequent decrease. For this to come about the microscopic constituents of the universe would not be started off moving about randomly after all, but each little particle, each electromagnetic wave, set off along a carefully chosen path to lead to this very special future evolution.

Consider what would happen in the latter half of such a strange universe. Instead of the stars converting hydrogen to helium and emitting radiation, the radiation would instead arrive out of deep space, carefully arranged to fall on to the surfaces of the hot stars. There it would percolate through the layers of the star, gradually combining together to form γ-rays, until each γ-ray arrived at the stellar core at the precise moment to split up a helium atom into its unstable constituents. In this situation, *hot* surfaces appear dull, and *cold* surfaces glow brightly. Instead of appearing dark, the sky would shine, as the cold depths of space gave up their remaining energy to the stars, which would appear as little black dots in the background glare, busily soaking up energy.

On the surface of a planet like the Earth, everything would run backwards. Rivers would flow uphill, raindrops rise into the sky and disappear, sand castles form on beaches under the action of the wind and the sea. Inanimate matter would form spontaneously into old and sick human beings who would grow younger and healthier, to end their days disappearing into mothers' wombs!

It is curious that this seems so laughable, because it is simply a description of our present world given in reversed-time language. Its occurrence is *no more remarkable* than what we at present experience – indeed it *is* what we actually experience – the difference in description being purely semantic and not physical. A human being in a reversed-time world would also have a reversed brain, reversed senses and presumably a reversed mind. He would remember the future and predict the past, though his language would not convey the same meaning of these words as it does to us. In all respects his world would appear to him the same as ours does to us – again, no more remarkable than our own.

What *is* remarkable, however, is the fact that our 'forward' time world *changes into* his 'backward' time world (or vice versa, as the situation is perfectly symmetric). Such a changeover requires, as we have seen, an extraordinary degree of cooperation between countless numbers of atoms.

When this idea was first mooted by Thomas Gold, he suggested that it take place in the context of a recontracting Friedmann model universe. In that model, even the cosmological expansion is reversed, so that both halves of the cycle would appear to their inhabitants identical, with 'normal' thermodynamics, retarded radiation and so on, as well as an expanding universe. Each set of inhabitants would regard their own half of the universe as the 'first' half – or expanding portion – and would assume that in the 'later' recontracting phase, the universe was 'really' collapsing, but its inhabitants would see everything going 'backwards', because thermodynamics and other asymmetries would be reversed. Naturally, neither set of inhabitants is correct in this assumption of their temporal priority. To regard one big bang as 'the beginning' and the other as 'the end' is mistaken. They are both in a sense 'beginnings' consistent with each other. For example, we could not know which half of such a universe we were living in now.

There is certainly something engaging about a model of the world with complete symmetry. The only trouble is, can such a possibility be consistent with what is known about physical systems? After all, in a time-symmetric universe, causes can come from the future as well as the past. Things can happen now, because someone millions of years in the future decides they should! Starlight emitted from the 'other half' of the universe, in our future, can reach us now, but 'backwards' in time appearing as advanced radiation rather than retarded. We could never see these stars of the future, then, because instead of their light falling on our eyes and stimulating the sensation of sight, just the opposite would occur. When we looked at such a star, our eyes would *emit* light towards it, rather than receive light from it. Whether or not we should notice such a phenomenon is unclear.

Not only could we not see this reversed world in our future, but we could not communicate with its inhabitants either. The reason for this is that they would be living, thinking and deducing 'backwards' relative to us. What to us constitutes information, to them appears as entropy.

The fact that radiation can travel between one half of this universe and the other makes it unlikely that the changeover would occur suddenly. John Wheeler has speculated on a gradual reversal – a 'turning of the tide' – in which time-asymmetric processes slowly run to a stop and then start to go backwards. If this were so, then certain minute signs might even now be discernible that at some epoch in the distant future the tide would turn. What is more, the idea has been taken seriously enough for at least one experiment to be performed to try to detect such a minute change, in the behaviour of radiation. The experiment (which failed to detect a tide on the turn) essentially amounted to a search for electromagnetic microwaves from the future.

The time-reversing universe is a fascinating intellectual curiosity, and a tribute to the fertile imagination of cosmologists, but should not perhaps be taken too seriously. However, the possibility that things can reverse themselves in time prompts the speculation that time itself might be cyclic. Until now, the topology of time assumed in these discussions has been that of the straight line, with a distinct *sequence* of events (though in which direction did not matter). Suppose that time had the topology of a circle instead – finite and closed, like the space of the recontracting Friedmann model? Such a topology would *force* a time-reversing universe.

The idea of a cyclic world is at least as old as Aristotle. In more recent years the general theory of relativity has yielded a number of situations in which the future histories of objects apparently join up with their pasts. It has never been clear just how physically meaningful such situations really are, but the implications of such possibilities for philosophy are profoundly disturbing. Free-will in a closed-time universe could not exist. The condition of a system could not be changed at will because its future would also be its past. So its present condition would depend on its *future* behaviour, which is what we seek to change!

If this type of universe were sufficiently complex, with a great number and variety of interactions, it is probable that we would not notice any such restrictions on the behaviour of physical systems. It would need to be a time-symmetric world, of course, so that it could 'get itself back' into its starting condition. It is doubtful if Gold's model interposes sufficiently complex interactions for drastic peculiarities to pass unnoticed. A better possibility would be to assume *two* cycles of expansion and recontraction. In one cycle, time asymmetry would be directed

Fig. 6.5. Time-reversing universe. In one cycle physical processes go one way, in the other cycle the opposite way. Time is closed into a loop.

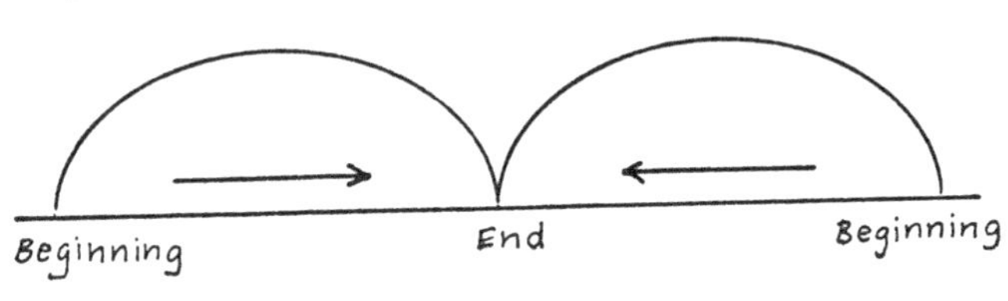

one way, and then reversed in the other cycle. There is no 'beginning' or 'end' to such a universe, but suppose we 'start' with our own big bang. The universe, at present in a state of expansion, would reach a maximum size and then recontract to a final bang where all structure and information would be wiped out. The matter content would emerge into the renewed cycle of expansion and recontraction with oppositely oriented time asymmetry – clocks running backwards relative to us. At the 'end' of this reversed-time cycle would be another big bang, which is then identified with the first-mentioned big bang in our own past. Nothing of any consequence could reach our part of the universe from the time-reversed part, but the accumulated starlight would appear in the big bang as radiation apparently present at what is normally interpreted as the 'creation' of the universe. Naïve calculations suggest this radiation to contribute a thermal background at this time of around three degrees absolute.

7 Mankind in the universe

7.1 The impact of space–time concepts on society

Mankind is a social animal, and scientists are no exception. The development of scientific theories takes place within the framework of a social and cultural order, comprising ethical, religious, economic and political components. The conceptual foundation for a scientific model of space, time and the universe is necessarily influenced by a pre-existing picture of mankind's place in the cosmos.

Conversely, experimental and theoretical advances in the scientific understanding of space–time physics and cosmology have an impact on society, as do all forms of human intellectual activity. These advances have not always been assimilated into the mainstream of knowledge with equanimity. Sometimes, the implications of new models of the universe have appeared so unpalatable that they have been fiercely resisted by the establishment, and even accompanied by violence, such as greeted the Copernican revolution.

Traditionally, people have appealed to religion to answer questions concerning the structure and evolution of the universe, the creation and destiny of all things in the great scheme. Frequently, scientific discoveries have conflicted with religious opinions, and science has come under attack in a variety of forms. One common criticism of the scientific explanation of these fundamental issues concerns its tentative nature. Religion is founded upon faith and dogma, so that the religious explanation is not expected to evolve in the light of experience. In contrast, science is ultimately empirical, and modifies its standpoint on the shifting sands of experimental and observational data, and therein lies much of its utility. However, it is not so that the continual re-arrangement of scientific opinions displays a sign of weakness. On the contrary, this is its strength. Science, like mankind, *evolves* to a more complex, yet a more powerful form.

It is, in fact, rather rare (in physical science at least) that a fully accepted theory is actually *wrong* in the strict sense of the word. Newtonian mechanics, and the associated model of space and time, served well for 200 years and more, and continue to serve well today. Newton was not wrong. The fact that his theory has been superseded by the theories of relativity and quantum mechanics means that the limits of validity of Newton's theory are now known. Both relativity and quantum mechanics *contain* Newtonian mechanics at the level of an approximation, which is an exceedingly good one in the everyday affairs of the world. Nobody would dream of using general relativity to compute the path of an aeroplane.

Science evolves better and better mathematical descriptions of nature, and society reflects this evolution in the changing perspective that new theories of space, time and cosmology bring to mankind's place in the universe. Perhaps this social impact is the greatest reason for humanity to continue research into these subjects. For thousands of years society was based on religion. During this time no universally satisfactory answers to the fundamental issues about the universe were given. Warfare, hatred and oppression resulted as religious groups sought to impose one particular set of beliefs upon other groups. In contrast, a science-based society has existed for very few years. In that time many of the burning questions so long pondered by the adherents of religion have been quietly answered. No wars, no hatred, no oppression have resulted between the proponents of scientific opinions, because science does not deal in beliefs, but in facts. A model of the universe does not require faith, but a telescope. If it is wrong, it is wrong.

The last few years have witnessed a collapse of confidence in science and the scientific explanation of nature. The simultaneous collapse of traditional religion has resulted in atavistic belief systems erupting in manifold grotesque guises. In place of reason, superstition is returning. Frequently, new cults pillage assortments of scientific concepts and weave them together in a pseudo-scientific, mystical mumbo-jumbo. The Western world has seen a resurgence of interest in witchcraft, flying saucers, spirit contacts and ESP. Exploiting the genuine peculiarities of some types of phenomena, cultists have forsaken rational explanation and imposed their own mythology on to a mixed hotch-potch of ideas. This incoherent belief system is then thrown in the face of the very science which has been plundered for their ideas.

The breakdown of scientific discipline, and the lurch towards medieval-style superstition, is undoubtedly due in part to the vagaries of technology. The confusion in the minds of many between science and technology has resulted in a backlash against the former on account of the shortcomings of the latter. Pollution, nuclear warfare, genetic engineering and mind control are all examples of the misuse of science in the form of technology. The superficial consumerism of capitalism, the alienation of people living in a world of push-button control, high-rise buildings and computer organisation, and the spoilation of the planet by energy-hungry industry, have all contributed to a revulsion against scientific values. Yet curiously, the same society, obsessed with the ethics of cost-effectiveness, decry all forms of scientific research which do not have immediate technological 'spin-off'.

Research into space, time and cosmology is the academic exercise *par excellence*. The general theory of relativity is probably the only major theory of science which has (yet) no conceivable application in technology. It is therefore a 'safe' subject. Occasionally the investigation of these subjects is justified on the grounds that science advances on a broad front. Research in one field tends to illuminate others with more immediate practical applications. In addition, new discoveries even in a purely academic area, can result in new technology. The classic example is Maxwell's electromagnetic theory. Devised as a purely mathematical unification of the properties of electricity and magnetism, it led directly to the prediction of electromagnetic waves and thereby to telecommunications, radio, etc.

Whilst this argument is certainly a valid one, in the author's opinion it is misplaced. The appropriate justification for academic research is not technology but knowledge. Mankind's understanding of the universe is the strongest motivation for the continuation of science. In our modern society based on mediocrity, knowledge is sacrificed to profit. Yet it is knowledge that most distinguishes mankind from cabbages. If society is not to reject science along with technology, the virtues of knowledge and understanding must be fully appreciated.

In a society with limited resources, the allocation of research effort in the correct order of priorities is always difficult. Is there really any reason to continue investigation of such an esoteric pursuit as the structure of space–time?

It is always easy to imagine that the job is finished. Just before the discovery of relativity and quantum theory, it was widely believed that the subject of physics was more or less exhausted. Existing theories appeared to account for just about all known phenomena, except for a few odd things that didn't at that time fit in. No one could know what great discoveries lay ahead, because none of the theories current at that time could foretell their own inapplicability. There is no reason to suppose, on examining Newtonian mechanics, that it will fail when applied to an atom.

The current theory of space–time is different in this respect. The general theory of relativity actually *predicts* its own breakdown; it contains the essence of its own limitations. This is manifested in the occurrence of the so-called singularities. These regions are boundaries of space–time and the theory of relativity cannot apply there. It follows that some new theory, a new model, is necessary. We may conclude that all of physics is not yet discovered. What this new theory will be like, can only be guessed. It may not even employ the concepts of space and time at all. It could be that a future society will not use these words or notions. Perhaps, like the ether, they will pass out of the interest and language of mankind. One thing is certain. We should be faint-hearted indeed if we ignored the challenge of the singularity.

7.2 Life in the universe

The development of our picture of space–time and cosmology in the last few hundred years has been accompanied by a changing conception of the place of mankind in the universe. In the pre-Copernican Western cultural bloc, mankind was endowed with the ultimate status of occupying the centre of all things. The Earth, mankind's purpose-built domain, stood like an axle around which the cosmic wheels revolved. All cosmic structure was subordinated to mankind's unique abode – the focus of all natural and supernatural activity.

The twentieth-century picture of mankind on the Earth could hardly be further removed from this egocentric obsession. Far from being a very special place, the Earth's position is now regarded as *typical* in many respects to that of all parts of the universe. With its attendant nine planets, the sun seems to be a very unexceptional type of star. Similar stars are scattered throughout the galaxy in their millions. Our galaxy seems to be a very unexceptional type of galaxy. Similar galaxies are scattered throughout the observable universe in their millions. If our sun and galaxy are so typical, then it is easy to imagine that our planet, biosphere and society are also typical features of the universe. Viewing the Earth in this cosmic perspective, it has become fashionable to regard *life* as one stage in the evolution of organisation in the universe. From the primeval fireball atoms were formed. In the stars which followed, complex nuclei evolved. The cooler regions around the stars witnessed the formation of still more complex molecules. Biological matter is a further step in the microscopic organisation of matter. According to this modern view, life arises in a natural fashion out of the raw materials provided by the stars. To imagine it is restricted to the Earth is an arrogant return to pre-Copernican egocentric dogma. Yet despite the fact that we know that distant regions of the universe have the same astronomy, the same physics and the same chemistry (because we can observe it with a wide range of hardware), the notion that other parts of the universe have the same biology still raises a considerable amount of controversy. The reason for this is partly due to the fact that not only has extra-terrestrial biology never been observed, but it is extremely difficult to go about observing it even if it exists.

If life really is a universal phenomenon, it drastically changes our whole assessment of mankind's place in the universe. The intellectual adjustment necessary is probably as great as the post-Copernican adjustment to the fact that the Earth is an insignificant speck. Is the living Earth also equally insignificant?

Let us examine some of the reasons offered for the pervasive nature of biology. Firstly, the foregoing remarks about the uniformity of physics and chemistry elsewhere enable speculation about life to proceed on the basis of terrestrial experience. If life on Earth is well understood, its position beyond the Earth may be inferred. What is required is a suitable extra-terrestrial environment for biological activity. Terrestrial biology depends on the stable thermodynamic disequilibrium produced by our proximity to a large entropy generator – the sun. Crudely speaking, we live in a temperature gradient. It is hard to envisage life existing under different circumstances. Astronomers can only observe distant matter because it is in disequilibrium. Indeed, it has been discussed at length in the earlier chapters that the entire universe is in pronounced disequilibrium, so there is no lack of temperature gradients around. In addition, there is the problem of stability. Life not only needs disequilibrium, it needs time. It has taken three billion years for biology on Earth to progress from a primitive sludge to a human being. This duration is a large fraction of the lifetime of the sun. Any minor change in the luminosity of the sun would have dire consequences for the delicate ecological balance which supports the more complex terrestrial life forms. Modern astronomy indicates that the sun is in a very stable condition. Although our vital requirement of the sun is the *disequilibrium* it produces in its vicinity due to the efflux of vast quantities of radiation, this efflux is an insignificant perturbation on its internal structure. The average photon of sunlight which strikes our eyes has taken about eight minutes to reach us from the solar surface. It has taken 100 million years to get to the surface from the centre. What this means is that the thermodynamic imbalance in the solar environment, so vital to the continuation of life, represents to the sun only a tiny leakage of energy from its periphery. There is thus no incompatibility here between thermodynamic disequilibrium and long-term stability. Although stars do pass through phases of violent activity and instability towards the ends of their careers, a large fraction of stars have, like the sun, been quietly radiating for billions of years, all of them possible candidates for the abode of life.

In addition to thermodynamic requirements, there are essential raw materials needed for life. Moreover, the delicate chemical processes necessary to initiate the spontaneous assembly of exceedingly complex organic molecules may well place very stringent restrictions on the type of environment suitable for biology. Much progress has been made in the post-war years by biochemists in understanding the physical and chemical conditions under which life will form. In 1953, a remarkable experiment was carried out at the University of Chicago by Stanley Miller and Harold Urey, in which the conditions thought to have prevailed on the Earth three or four billion years ago were simulated in the laboratory. At the end of the experiment (which lasted several days) large quantities of important organic molecules had been synthesised. While far from producing *living* matter, this experiment, and the many which have followed, have supported the contention that under a fairly wide range of conditions, large quantities of 'pre-biological' molecular building blocks rapidly form. The significance of this discovery hinges on the fact that all terrestrial life, from a bacterium to mankind, is composed out of combinations of a small number of such building blocks. It may be hard to arrange a laboratory experiment to produce spontaneously even the lowliest living organism in a week or even a decade, but given a few million years, it is considered by many biochemists virtually certain that such an event will occur.

There is a sense in which the transition from inanimate building blocks to the first self-replicating, respectable, living thing is of a much greater consequence biologically than all the subsequent evolution of the primitive first organisms into the vast range of sophisticated bioforms which now inhabit the Earth's surface. The weak link in the chain is the first step, and the status of this step is still far from definitive. Nevertheless, following for the moment upon the optimism of the biochemists, it is concluded that most stars of the same general type as our sun, if they possess planets of the same general character as the Earth, would evolve life. Unfortunately, there is no observational check available on the existence of Earth-like planets outside the solar system. The Earth itself is far too small to be seen through a telescope of reasonable size even from the nearest star, and conversely our terrestrial telescopes see nothing of small planetary bodies in other systems. However, within our own solar system there are other similar planets (Mars, Venus), and theories of planetary formation support the conjecture that most stars possess similar bodies. Planetary bodies unlike the Earth (much larger) have been discovered around some nearby stars, and some biologists have speculated that life might form under the very different conditions which prevail on them. Life as we know it is based on carbon, and probably requires large quantities of water, but it is conceivable that alternative biologies could develop from an entirely different chemical base. Actually, such speculation is of little consequence to the overall problem. The monumental question is whether life is extant throughout the universe, or is a 'miraculous' accident in our own little corner. Alternative biochemistry only contributes a small factor to load the odds in favour of the former.

The upshot of the recent developments in understanding the chemical basis of life is the emerging view that biological substance is a kind of alternative physical state of matter – gaseous, liquid, solid, biological – with the implication that its formation proceeds naturally and automatically under the right conditions. The American astronomer Carl Sagan has written 'The origin of life on suitable planets seems written into the chemistry of the universe.' In truth, we simply do not know at this time what the chances are of life elsewhere in the universe, but perhaps on general grounds we may indulge in some cautious optimism that inhabited planets may be rather a common occurrence.

Founded upon this assumption has arisen the nascent science of exobiology – the study of life beyond the Earth. To date it has no subject matter but plenty of theory! Two main experimental procedures suggest themselves for the discovery of extraterrestrial life. The more straightforward method is also generally the less hopeful. This is direct spaceflight. The spectacular successes of the space programmes have drawn much attention to the possibility of travel to other worlds, with perhaps even the hope of encountering other life forms. Certainly, restricting attention to our solar system of nine planets, it is a reasonable assumption that future technology will enable that possibility to be realised. The chances of encountering life on these sister planets is slender, but not impossible. What is known of the conditions of Mars (and possibly Jupiter) does not encourage the speculation that life exists there, but is not inconsistent with it either. A few scientists are even of the opinion that very primitive organisms are fairly likely on at least one of these planets, and there is no doubt that even a single bacterium found on Mars is worth a thousand speculations. It would be a truly profound discovery in our continually changing perspective of the universe.

However, if no extra-terrestrial life is found in the solar system, it is not simply a matter of building bigger and better rockets to travel to the stars. The *nearest* star is four and a quarter light years away (compared with one and a quarter light seconds to the moon). At currently available rocket speeds it would take thousands of years for an explorer to reach it. Greater speeds will certainly be available to a future generation. If speeds approaching that of light were achieved, the time dilation effect would reduce the travel time for the crew, and in principle this would enable voyages over thousands of light years – right across the galaxy – in a single human lifetime. However, due to the 'twins effect', the spacecraft would return to Earth thousands of years in the future, to find that the society which dispatched it had long since disappeared. Quite apart from the technical difficulties of developing near-light velocity space transport systems, the energy expenditure required to reach these speeds is colossal. A typical light spacecraft might require as much as a million million tons of fuel to achieve 99 % of light speed.

Future technology may well invent new propulsion mechanisms (some have been proposed), but there are basic reasons why any mode of interstellar transport, even if feasible at all, would make huge demands upon our global resources. Such a venture might be undertaken if sufficient motivation could be found – the certainty of contact with another intelligent civilisation for example.

This is really the rub. The technological problems (though formidable) might be solved, but no one is going to mount such a gigantic technological exercise without a very good reason. If extra-terrestrial contact is considered good enough, we are faced with the daunting task of where to find it. Even if life is a universal phenomenon, only exceedingly crude estimates may be made of how *common* it will be. Moreover, there is no agreed way of assessing the ubiquity of *intelligent* life.

Making the optimistic assumption that intelligent beings will automatically evolve on biologically viable planets, it has been estimated that the number of intelligent communities in our galaxy is about 10 times the average lifetime of these communities measured in years. The latter quantity is of course quite unknown, and depends to some extent on what we mean by intelligence.

Human society can be credited with a few thousand years of civilisation, which might be nearing its own destruction through technology. If that is so, and our experience is typical, there are perhaps tens of thousands of planets in the galaxy with intelligent civilisations. On the other hand, it could be that the lifetime of civilised communities should be measured in *millions* of years or more. If so, there may be tens or even hundreds of millions of inhabited planets in the galaxy.

This all seems very impressive, but the problem of where to look is still a formidable one, because the galaxy contains *100 billion stars*. Even 100 million likely abodes for intelligent life would require us to sample literally *thousands* of planets before we had a reasonable likelihood of success. It would demand journeys to all the likely stars within 100 light years or so. It is hard to resist the conclusion that physical contact between planetary civilisations is an exceedingly rare occurrence in the universe, on purely motivational grounds. (Naturally, this conclusion is subject to the caution that the motivations of alien civilisations perhaps millions of years in advance of our own are quite unknown, and may be incomprehensible to us.)

Perhaps the strongest argument against the likelihood of interstellar travel is its pointlessness. Terrestrial exploration has always been either for colonisation, trade or information. The first two can be ruled out for the interstellar case. The idea of transporting entire populations or consumer goods across light years of space is ridiculous. In the cosmic context information is by far the most important commodity to exchange between civilised communites. However, it is not necessary physically to travel to distant stars to exchange such information. This can be done by radio, for example. Nothing can travel faster than electromagnetic waves, so that in terms of time this is the most efficient communication channel of all. Once again, we are faced with the problem of where to look. The fraction of civilised societies in the galaxy which have evolved a technical ability to communicate by radio might be quite small, so that the location problem becomes even worse. Nevertheless, a moderate-sized radio telescope can systematically scan thousands of likely sights in the hope of receiving a message of some sort. The largest radio telescope in the world, at Arecibo in Puerto Rico, is capable of communicating with a similar device anywhere in the galaxy.

In recent years a number of attempts have been made to detect radio signals from technical civilisations nearby in the galaxy, so far without success. In addition, messages have been transmitted from Earth. Although the entire idea may be a waste of time and money, it certainly seems worth a modest effort to try and establish such communication in view of its enormous significance. The reader who is concerned that signalling our presence to a possibly bellicose alien species might be somewhat foolhardy need not worry too much. Even at the speed of light, radio waves would take 100 years to reach a civilisation 100 light years away. We could not receive a reply, let alone an invasion fleet, for at least 200 years.

Because the human race has already embarked upon rudimentary communication attempts, one thing should be appreciated. If such attempts have any likelihood of success, it is because the average lifetime of technical civilisations is millions of years. As our own technological society is only a few decades old, we will be the *youngest* such society in the galaxy. A responding community would almost certainly be incomparably, and perhaps incomprehensibly, more developed scientifically, culturally and ethically than ourselves. Indeed, the most intelligent products of their community might not be biological at all, but machine intelligence. Even at our own level, a large part of any discourse would be devised, analysed and executed by computers.

Knowledge from the scientific future could endanger life on Earth still more, but a million-year-old civilisation would need to have solved its own social problems. Perhaps information for a new technology would only be transmitted to us after information for building a new society.

7.3 How special is the universe?

It is a striking thought that 10 years of radio astronomy have taught humanity more about the creation and organisation of the universe than thousands of years of religion and philosophy. It is of some interest to examine how recent advances in astronomy, physics and cosmology have contributed to the scientific picture of mankind in the universe, and to compare this picture with traditional religious beliefs.

Mankind traditionally regarded the universe as purpose-built. Things are organised the way they are for the convenience of human life. Our environment abounds with facilities for making life easy and pleasant. There is plenty of water to drink and air to breathe. Harmful radiations from space are excluded by the atmosphere. The sun lights and warms our days, but conveniently lets us sleep at night. It radiates at just the right temperature to keep us comfortable, and doesn't fluctuate. Geological catastrophes occasionally occur, but not in England. Isn't it all too good to be true?

It is very difficult to determine just how delicately life is balanced on the scales of physics and chemistry. Life has *evolved* on this planet, and thereby adapted itself to the prevailing conditions. Rather than our world being made comfortable for us, we have been made suitable for it. Just how far the organisation of the universe could be changed before all conceivable life forms are impossible is uncertain. It is frequently claimed that minor changes in a few of the apparently arbitrary constants of nature, such as the strength of the nuclear coupling, would imply drastic changes in the conditions in the universe. For example, free hydrogen – the solar fuel, and mainstay of life on Earth – would be rapidly synthesised entirely to helium in the big bang if this coupling were a very few per cent stronger. However, in the absence of a suitable knowledge of living matter under wide ranges of conditions, one should be cautious in drawing definitive conclusions about how precarious biology really is in this universe.

Two opposing views may be expressed with regard to our own existence. One, that the universe has been created in a very special way to enable life and humanity to evolve. The second, that if things were not the way they are, we would not be here to wonder about it. Both of these views are consistent with the notion that the presence of life *constrains* the universe, to a greater or lesser extent, to possess certain features. It is interesting to examine just what aspects of nature *must* be observed by virtue of the fact that we are here to do the observing in the first place. From time to time, our very existence is indeed offered by some scientists as a *consistency* with certain features of the universe.

As a first example of this reasoning, recall that the contents of this book have dealt at length with the topology, geometry and asymmetry of the universe, but no mention has been made of its *size*. The sheer enormity of the cosmos always inspires awe. Billions of stars are spaced out by vast distances measured in light years, galaxies are separated by millions of light years. To help in visualising the scale of things, imagine that the Earth's orbit around the sun, which is nearly 300 million kilometres across, was reduced to the size of a penny with a small speck at the centre for the sun. The nearest star would be about two kilometres away. The galaxy would be so large it would cover the surface of the Earth. The Andromeda galaxy, the only one barely visible to the naked eye from Earth – would be a half a million kilometres away – about where the moon is. The farthest galaxies visible through our most powerful telescopes would be a billion kilometres away. Indeed, the density of the universe is so low that there is an average of only one atom for every 1000 litres of space. If all the matter in the universe were concentrated into blobs with the density of water, these blobs would only occupy 10^{-28} of 1% of the available space!

Why is the universe so big?

First remember that the universe is not static in this condition, but expanding. There is good evidence that in the past it was in a very dense state. The expansion is necessary to prevent it falling in on itself to a singularity. The expansion is (probably) slowly decelerating at a rate determined by the density of matter which is encouraging it to fall back. In the Friedmann models described in chapter 5 this determination is unique. That is to say, the deceleration rate of the universe is fixed by the density of matter. If the universe were much denser, the deceleration would be much faster.

It follows that the present observed density of stars is fixed by the *age* of the universe. There is no simple Friedmann model which allows a 10-billion-year-old universe to have, say, stars spaced out by a few light days.

This is where the biology comes in. In order to evolve intelligent beings (us) a biological system needs billions of years. Evolution is an exceedingly gradual process, requiring an enormous amount of false starts. It is dependent upon vast numbers of small accidents from one generation to the next.

In addition, life on Earth (and probably any life) is based on carbon. This carbon was synthesised in heavy stars a few billion years ago. It takes millions of years for these massive stars to form, synthesise the carbon, and explode. Either way, if the universe were substantially younger than billions of years, we could not be here to see it. The universe is so big because it is so old. It follows that our own existence constrains the stars to be very, very far apart. It is ironical that the same conditions which are necessary for intelligent life to form are also responsible for preventing physical contact between these intelligences.

There is another sense in which the universe is large, to do with the sheer numbers of stars in the sky. Actually, a casual glance at the night sky gives the impression of millions of stars, but that is false. A person with average eyesight can at most see a few thousand. However, within the range of current optical telescopes are some billion billion stars. Adding up all the atoms in each gives the unnervingly large number of 10^{80}. Why is this number so big?

The size of the universe in this context is somewhat ambiguous. In the ever-expanding Friedmann models, the volume of space is infinite, which implies an infinity of stars extending outwards for ever in all directions. However, we do not have visual access to all these stars. Even if we live in a universe like that of the recontracting model, which has a finite volume, only a fraction of that volume would be visible to us at this time, however powerful our telescopes. The reason for this is that in a universe 10 billion years old it is only possible to see a distance of 10 billion light years. Beyond that corresponds to 'before time'. The limit of visibility is the horizon referred to on page 162. The older the universe, the further away the horizon – it recedes from us at the speed of light. It follows that the large number of stars in the universe is due to the great distance of the horizon, which in turn depends on the great age of the universe. Once again, this large number need occasion no surprise. The presence of cosmologists requires it.

A similar basic question to the foregoing is 'why is the universe so dark?' In chapter 5 this question was answered in the context of Olbers' paradox. However, that was not really the whole story, because it dealt only with starlight. The universe began with a hot bang, and has since cooled down, because of the expansion, to a mere three degrees or so above absolute zero. The sky is therefore not completely black, but glows dully in the far infra-red region of the spectrum. Special radio telescopes are needed to detect this primordial 'space-glow'.

There seems to be no reason why the present three degrees absolute should not be 300 degrees absolute (around room temperature). However, if it were, we could not exist. Firstly, because that is about the temperature of the Earth, so the crucial thermodynamic disequilibrium necessary for life could only be established on a much hotter planet on which water evaporates. As water is probably also crucial for life there is obviously a problem here. Secondly, and more seriously, such a high radiation level would prevent the formation of galaxies by dominating over matter with its gravitational attraction. Life could not form without galaxies.

Several other attempts to check the consistency of intelligent life with fundamental features of the cosmos have been made. The British mathematician Brandon Carter has addressed the question 'why is gravity so weak?' Recall that gravity, 10^{40} times more feeble than electric forces in an atom, nevertheless rules over the motion of the universe. By considering the evolution of stars, Carter has shown that this ratio of forces determines the lifetime of stars; and long-lived, stable stars are an essential prerequisite for intelligent life.

In a different vein, Stephen Hawking and Barry Collins took up one of the most mysterious unanswered questions of all: 'why is the universe so isotropic?' This problem is being tackled in several ways, and was briefly alluded to in section 5.5. Hawking and Collins suggest that only in an isotropic universe will galaxies form and life evolve. Their reasoning is based upon an examination of the initial conditions on the large-scale motion of the universe which are required for the subsequent observed isotropy.

As a final example, attention has already been drawn to the necessity of a thermodynamic disequilibrium in the universe in order that biological systems may exist. The time asymmetry of the world, so conspicuous in everyday life, is an indispensable ingredient for that life.

This by no means exhausts the possible list of these bio-cosmological considerations. Other fundamental features of space–time, such as dimensionality, may well be amenable to this type of investigation. It is important to realise that the existence of intelligent life in the universe does not *explain* these features. It merely indicates that if they were very different, we should not be here to know about it. As mentioned in chapter 6, some cosmologists have suggested that there is not one universe, but many, each one realising a different set of conditions, and perhaps, laws of physics. The reason we happen to have picked out such a very special one (big, isotropic, cold, etc.) to live in is simply because it is the only type of universe we *could* live in.

Although the large-scale structure of the universe – its size, the distribution and disequilibrium of matter – appear to be constrained by the existence of cosmologists, the small-scale structure has the opposite status. The traditional religious view is that the local structure, the Earth and its surface features, the sun, etc., represent a very special organisation of the world. This highly distinctive order is supposed to have been built into the world at the creation. In contrast, the small-scale systems of stars and planets are regarded by modern science as forming *naturally* and automatically out of the primeval fireball. Indeed, rather than require that the creation contain this structure at the outset, precisely the contrary appears to be necessary. The universe began in local equilibrium with the microscopic motions randomly arranged. The beginning was a state of disorder. Organised structure has arisen from disorder quite automatically as a result of the cosmological expansion. The microscopic condition of the universe at the time of the creation need only be a completely random one. It is no longer necessary to assume that the organisation of the world requires an organising agency to create it in a special condition. Such organisation follows naturally from the laws of physics and the expansion of the universe, under the widest imaginable range of small scale initial conditions.

The scientific picture which emerges is thus a curious inversion. Far from attributing the character of our immediate environment, including our own existence, to miraculous local events, while regarding the large-scale structure of the universe as irrelevant, it now seems that it is *cosmology* which is the significant factor, the local situation taking care of itself. Provided the global properties of the universe are appropriate, it seems almost inevitable that stars and planets, life and intelligence, will appear throughout the cosmos.

How special is the universe then? Globally, very special, but locally quite unremarkable.

Such a perverse conclusion may well appear unpalatable to the reader, who is expected to thank for his or her existence the distribution of matter and radiation in the remote regions and early moments of the universe, rather than a purpose-built planet Earth. Quite apart from whether life is a universal phenomenon, when viewed from this perspective, the arrival of mankind in the universe is a cosmic event.

We have come a long way from the biblical account of the creation. In the bible, light and warmth, organisation and life, arose out of darkness and the void. The universe is accepted as an *effect*, caused by the *action* of God responding to a *prior* motivation to build form into a pre-existing but uninteresting space and time. The modern scientific account is a strong contrast. The universe began in searing heat and light, and has cooled and darkened. The evangelist's 'Let there be light!' receives the scientist's response 'Let there be darkness' for it is only in a dark, cold universe that the energy locked in the sun can be usefully made available to drive living systems on Earth. Moreover, space and time themselves are considered as *physical entities* by the physicist. Einstein's general theory of relativity demonstrates how the explosive appearance or disappearance of matter occurs at a *boundary* of space–time. If the universe really was created 10 billion years ago, and is not infinitely old, then space–time too came into existence then. The initial singularity is truly an effect without a prior cause, for there is no pre-existing space or time – or anything physical at all – to contain this cause. To imagine God reigning in an earlier phase of the cosmos, and being motivated to *cause* the universe, is quite misguided, and the result of attributing to the Deity an over-anthropomorphic status. Not only do notions like cause and effect require the existence of *time* in which to operate, but also they require time *asymmetry*; but time, and especially the asymmetry, are concrete properties of the *material* world, having meaning only *after* its creation, and indeed, in the latter case well after, when the primordial equilibrium had been removed by the cosmological expansion.

There has been a pattern of progress in the history of mankind's attempt to invoke supernatural agencies to account for characteristics of natural phenomena, the causes of which are obscure. Primitive communities, with no knowledge of physical science, invoked gods of all varieties and dispositions to 'cause' rain, flooding, lightning, comets, etc. These early gods were strongly anthropomorphic in character, widely believed to possess material bodies of human physiological appearance, and intellectual capabilities and motivations scarcely different from that of children. In short, superhuman humans. Even early attempts at monotheism could not resist the temptation to create God in man's image as a somewhat fickle supernatural warrior, engaging enthusiastically in the feuds of local tribes.

With the growth of physical science, the Renaissance and the age of enlightenment, supernatural agencies were gradually banished from physics and astronomy. The prospects for a physical individual with the body of a human but the power of the Almighty became extremely dim. In the words of the British theologian John Robinson, the God 'up there' became the God 'out there'. The science of astronomy did not leave room for a physical heaven in the sky, and a new conception of God as a non-material entity, transcending the concrete world of space and matter, arose and flourished; a God beyond matter.

In spite of the staggering successes of the physical sciences in explaining natural phenomena without supernatural causation, ignorance of biological and social systems left plenty of scope for supernatural involvement in these areas. God may not have been necessary to explain the motion of the planets, but He was still indispensable for the creation of life. The Darwinian revolution displaced God back in time by three billion years, just as the astronomical revolution had displaced Him out of space. Mankind was not miracle, but the product of evolution – accident and brutal survival – over aeons of adaptation from the simplest living thing. The gradual unravelling of the chemical and physical basis of life is just an inevitable, further step in the explanation of the physical world in terms of scientific principles. Although laboratory experiments do not have the millions of years necessary to produce truly living matter out of inanimate elements, the building blocks of life have been made, and the simplest living thing has been dismembered into these building blocks. The creation of life is no longer a mystery. We have moved on to the next step: a God beyond life. Some would even maintain that our understanding of social and ethical organisation now displaces God entirely from the worldly affairs of men.

In hindsight then, the idea of appealing to an act of God only at the cosmological creation seems rather desperate. Mankind's egocentricity in creating an anthropomorphic Deity, has led to the steady displacement of such a Deity from all things of relevance in the material world. To attribute the creation of the universe, even if there was one, to an act of God, is falling into the same trap as assuming a God of matter or a God of life.

It is assuming for God the status of a human once again, a human, moreover, existing *in* the created world, with its time asymmetry and temporal order of cause and effect.

It has been emphasised several times in this book how fundamental physical time is in giving expression to our whole conception of mankind and the universe. The distinction between past and future permeates our whole being. The remembered past is regarded with nostalgia, or remorse. The future is faced with fear or hope. All human actions are framed by past experience and future expectation. *Motivation* is likewise a product of time asymmetry. It follows that to attribute motivation to God in the absence of time asymmetry, and even of space, time and matter is absurdly anthropomorphic. As already emphasised, the creation of the universe cannot be caused by a prior motivation; that is logically contradictory. It is necessary to move on to a still more radical concept; a God beyond space–time.

Is it possible to attribute the creation to events which occur *after* it, in the fashion of the advanced effects of the Wheeler–Feynman absorber theory? What then of the time-symmetric universes, say, those which collapse to a 'negative creation', or end; how can both temporal extremities be 'caused' by what happens in between them?

The best response to these questions is a denial of the relevance of cause and effect. They are essentially human concepts for human situations. At best, in the world of physics they describe time-oriented interactions in terms of the decay of organisation, itself (as remarked on page 81) a purely human notion. Far better to regard the universe as a *total phenomenon*: in the words of the German mathematician Hermann Weyl (1885–1955) 'the world doesn't happen, it simply is'. The world need not be started off, to run its carefully arranged course to some unknown destiny. Rather the world *is* space–time, matter and interactions, extending from past to future, from place to place, from event to event in a vast network of complexity and existence.

7.4 Mind in the universe

The cosmic perspective may be illuminating and awe-inspiring, but it is not the perspective of mankind. The individual looks out into the universe around and perceives, interprets, conceptualises. Unlike a God beyond space–time, he or she is part of space–time. In this sense, the theories of the universe discussed in this book don't mesh with the actual experience of human beings. This mis-match is because human beings don't see the *total phenomenon*, but view the universe through a little window – the window of our minds.

The picture which we see through the window is a 'movie'. It moves. The world is full of *activity*. Why?

Things happen because time *passes*. What more obvious statement could be made? Yet how utterly incomprehensible it turns out to be. How can time move? Time is part of space–time, what can time move *in*? How fast does it move? One day per day!

The impression of a moving, flowing, passing time, a time of uni-directional activity, is so fundamental to all experience (at least in Western civilisation) that it pervades our entire society. The reluctance to discard the passage of time as an illusion is tremendous.

Human psychological time possesses many levels of structure, beyond the time which enters into physics. Physics distinguishes between past and future, but mind distinguishes past, *present* and future. We *remember* the past, we *plan* the future, but we *act now*. The present moment is our moment of access to the universe – we can always change the world at this instant.

But what is now? There is no such thing in physics; it is not even clear that 'now' can ever be described, let alone explained, in terms of physics. For example, suppose the following is tried. 'Now' is a single instant of time. The response 'which instant?' yields the answer 'every instant'. Each instant of time becomes 'now' when 'it happens'. But this is going round in circles. At the time of writing the year 2000 is in the future. One day (2001) it will be in the past. Although it is in the future 'now', it will 'occur' when 'now' is the year 2000. To counter that all time is 'now' eventually, but not all at the *same* time is mere tautology. It simply describes one-dimensional time as a collection of 'nows' instead of a collection of points: a minor semantic re-arrangement. Notions such as 'the past', 'the present' and 'the future' seem to be more linguistic than physical.

In the face of this impasse, physical science can make very little progress in the discussion of the now. However, the special theory of relativity does throw some light on the issue. Recall from section 2.2 that simultaneity is *relative*. There is no universal present moment at every point of space. Events occurring at spatial separations which cannot be connected by a light signal cannot be assigned a particular chronological order which is the same for all observers in all states of motion. So one of the characteristics of the mental 'now' – that all people everywhere are experiencing the *same* now (e.g. I wonder what so and so is doing now) – is an unjustified extrapolation. There is no universal now, but only a personal one – a 'here and now'. This strongly suggests that we look to the mind, rather than to the physical world, as the origin of the division of time into past, present and future.

We still have to contend with the fact that the now of our conscious awareness appears to *move* steadily from past to future. It is this motion, rather than the asymmetry of memory and prediction, which seems to create the strong mental distinction between past and future. Much language and metaphor incorporates this sensation: time 'flies', things 'come to pass', a person 'passes away'. Human beings have a strong sense of the future 'coming into being' while the past has 'passed out' of existence (e.g. 'that's over with'). Only the present 'exists'. There is a sort of mental continuous creation – a new world every moment. The interconnection of successive worlds then gives the impression of one 'changing into' or 'passing on to' the 'next'.

None of this appears in physics. No physical experiment has ever been performed to detect the passage of time. As soon as the objective world of reality is considered, the passage of time disappears like a ghost in the night.

A few years ago, some very unusual books were written by J. W. Dunn on the notion of serial time. Based on the idea that a moving time must be measured against another time to make sense (e.g. how *fast* does time move?) Dunn proposed an unending series of time dimensions, each one engaging in a flux relative to the next. In support of this extraordinary proposal the author cited examples of apparent precognitive experiences associated with dream states. Although Dunn's ideas have not been generally accepted by physicists, it is curious that the latter always seem very reluctant to deny the reality of a moving present. Unless this moving-now phenomenon is to be discarded entirely, it has to be admitted that we do not understand something about time, or about the mind, or both.

In the emerging picture of mankind in the universe, the future (if it exists) will surely entail discoveries about space and time which will open up whole new perspectives in the relationship between mankind, mind and the universe.

INDEX